DISMANTLING THE MEMORY MACHINE

SYNTHESE LIBRARY

STUDIES IN EPISTEMOLOGY,

LOGIC, METHODOLOGY, AND PHILOSOPHY OF SCIENCE

VOLUME 128

HOWARD ALEXANDER BURSEN

DISMANTLING THE MEMORY MACHINE

A Philosophical Investigation of Machine Theories of Memory

D. REIDEL PUBLISHING COMPANY

DORDRECHT : HOLLAND / BOSTON : U.S.A.

LONDON : ENGLAND

Library of Congress Cataloging in Publication Data

Bursen, Howard Alexander, 1949–
Dismantling the memory machine.

(Synthese library)
Bibliography: p.
Includes index.
1. Memory. 2. Neuropsychology. I. Title.
BF371.B86 153.1′2 78–10492
ISBN 90–277–0933–5

Published by D. Reidel Publishing Company,
P.O. Box 17, Dordrecht, Holland

Sold and distributed in the U.S.A., Canada, and Mexico
by D. Reidel Publishing Company, Inc.
Lincoln Building, 160 Old Derby Street, Hingham,
Mass. 02043, U.S.A.

Printed in The Netherlands

To Norman Malcolm and Bruce Goldberg

ACKNOWLEDGEMENTS

I wish to acknowledge the debt I owe to those friends with whom I have discussed – and argued – these matters over the past several years. I especially want to thank James Gregory and David Oppenheim; their astute criticisms helped keep my thoughts in an orderly path. Special thanks are due to Raymond Jaffe, whose suggestions helped to prepare this work for publication.

I wish also to thank the following editors and publishers, for their kind permission to quote from works prepared by them:

Dr. D. A. Norman, editor of *Models of Human Memory,* Academic Press, 1970.

Dr. D. E. Broadbent and Dr. K. H. Pribram, editors of *Biology of Memory,* Academic Press, 1970.

Dr. William S. Fields, editor of *Information Storage and Neural Control,* Houston Neurological Society, 1963. (Also edited by W. Abbott).

Dr. K. H. Pribram, editor of *Brain and Behaviour 3: Memory Mechanisms,* Penguin Books, 1969.

TABLE OF CONTENTS

PREFACE

The subject of the following study is theories of memory. The first part is a study of one broad type of theory which is very widely adhered to at this time. It enjoys great popularity among neurophysiologists, neuropsychologists, and, more generally, among scientifically oriented people who have directed their attention to questions about memory. Further, this way of looking at the matter is not confined to scientific professionals. Indeed, we can find popularized versions of the view in magazines like *Time* and *Reader's Digest*. So in the first part of the book, I will give a presentation of the view in its general form. The theory will be presented in such a way as to reveal the features which make it tempting, which make it seem to be a very natural way to explain the phenomena of memory. (And, clearly, from the number of adherents the view has won, it *is* tempting, and it *does* seem to be a natural way to go about explaining memory.) After setting forth this generalized version of the theory, I will next present material by various authors who hold this view. This will allow the reader to get some idea of the different forms which the theory (the 'memory trace' or 'engram' theory) takes. The last step is a criticism of the theory.

In the second part of the book, the attack on trace theory will be strengthened by a further criticism. The second part will conclude with an examination of two very powerful variants of trace theory. Both the 'stimulus-response' and 'information processing computer' theories of memory are current in today's scientific community. In the second chapter, we examine them and show that they, too, fall prey to the criticisms brought to bear against other versions of trace theory. This comes about because the criticisms are general, and do not depend upon the particular form the theory takes. It is the requirement of the trace itself which exposes all trace theories to these criticisms.

If the criticisms here brought out against trace theory are justified, the consequences for current memory research are of direct practical import. Not to mince words, if what is maintained here is correct, then many theorists have written in vain, and much neuropsychological research is entirely misguided. It is, perhaps, surprising that a challenge of this sort should appear in a book of philosophy, written by one untrained in neuropsychology. The reader will surely wonder how neurobiological memory mechanisms could possibly be a subject of *philosophical* inquiry. The final third of the book deals with this question. Briefly, my claim is this: While trace theory certainly *seems* to be scientific, it is not. Indeed, trace theory is nothing but a disguised *philosophical* theory. There are three considerations which support the claim that trace theory is philosophy, not science. First, it is historically an ancient theory put forward by philosophers. And the latest versions put forth by scientists are, underneath their technical jargon, the same old theory in new dress. Second, the criticisms here levelled against the theory are philosophical criticisms (and not, for example, scientific ones). And finally, the core of the theory itself – I mean the very notion of the memory trace – is a philosophical notion. That philosophical core is exposed and examined in the last section, and is found to be unsound.

Naturally, machine explanations of human memory are intimately linked with machine explanations of all sorts of human abilities (speech, for example). And, of course, such explanations go hand-in-hand with the view that people are, indeed, nothing but incredibly complex machines. So now we see that an attack on a theory of memory can lead to a much larger problem. An issue of such overwhelming importance (whether or not people are machines) is not to be settled in the final third of a book on machine theories of memory. A thorough treatment of that topic would take us too far afield from the subject of memory theories. But the importance of the issue will serve as justification for making a short remark now. This book purports to show not only that mechanistic accounts of memory are inherently unsatisfactory, but also that the mechanistic assumptions behind such theories lead to a paradoxical conclusion. The paradox is this: If one is a mechanist, then one is forced to hold that the human brain has magical or supernatural powers. And, while I myself do not believe that the brain is magical or mystical, I believe that any mechanistic theorist is committed (unwillingly, no doubt) to this position. If so, the view that people are machines is not science, but pseudoscientific nonsense. That this nonsense arises from a philosophical error, I hope to show in the course of our examination of trace theory.

PART I

THE TRACE THEORY OF MEMORY

CHAPTER ONE

AN INTRODUCTION TO TRACE THEORY

One good way to gain an understanding of trace theory is to see what sort of question it purports to answer. An example of this sort would be: "When a person hums a tune he is familiar with, how can he possibly get it all right?" Trace theory is an attempt at providing a scientific answer to this question, and, of course, to a great many more like it. Other questions of this type might be, for example, "How can a person recognize a friend walking down the street?", or, "How can I recognize the Mona Lisa, without confusing it with some other painting?", or, "When I say that a drink smells like strawberries, how do I pick *that* smell, rather than (say) rotten eggs?"

These questions are to be answered in the following way. The brain (or perhaps the entire central nervous system) is nothing other than a super-powerful, ultra-sophisticated computer, hooked up to incredibly sensitive microphones (ears), cameras (eyes), and so on, including all the organs of sense, and all the sensitive areas of skin, etc. Now in the case of someone humming a tune he or she is familiar with, the microphones of the ears are hooked up to a very high fidelity recorder. Naturally there are no reels of magnetic tape, or anything that crude, but nevertheless there must be some biological mechanism which works in the same way, laying down a recording, a *trace* or *engram* of the piece of music in question. Dr. Thomas A. Harris, author of the popular book, *I'm O.K. – You're O.K.*[1], says,

One noted explorer in this field is Dr. Wilder Penfield, a neurosurgeon from McGill University in Montreal . . .
. . . Penfield conducted a series of experiments during which he touched the temporal cortex of the brain of the patient with a weak electric current

> transmitted through a galvanic probe.
> The evidence seems to indicate that everything which has been in our conscious awareness is recorded in detail and stored in the brain and is capable of being *played back* (his italics) in the present (p. 5, *op. cit.*)
> Another conclusion we may make from these findings is that the brain functions as a high-fidelity recorder, putting on tape, as it were, every experience from the time of birth, possibly even before birth. (The process of information storage in the brain is undoubtedly a chemical process, involving data reduction and coding, which is not fully understood. Perhaps oversimple, the tape recorder analogy nevertheless has proved useful in explaining the memory process. The important point is that, however the recording is done, the playback is high-fidelity.) (p. 9)

Quoting Penfield, the author says: " 'Whenever a normal person is paying conscious attention to something, he simultaneously is recording it in the temporal cortex of each hemisphere.' " (p. 9, *op. cit.*)

One gets the feeling that the trace theory, far from being a theoretical hypothesis about how memory is possible, is actually a well-confirmed fact, something like the fact that the heart functions as a device which pumps blood. It seems as if the only questions left are questions about just how the recording device operates. And though there are a number of competing theories about the memory trace, it seems unthinkable to deny that some sort of trace theory is correct. To deny this would, perhaps, be something like joining the Flat Earth Society. In the past fifteen years or so, many books have been written, and many symposia have been held on the subject of memory theory and research. In all of these writings and researches, it is assumed that some sort of trace theory must be correct. Books with titles like *Brain and Behavior, Models of Human Memory*, are not questioning the existence of memory traces. They are only seeking to find out just what sort of traces there are. Wolfgang Köhler, in his enormously influential book, *Gestalt Psychology*[2] says,

> As to questions of detail, various hypotheses may be invented. But no theory will be acceptable which fails to assume the existence of some trace. (p. 149 of the paperback edition)

Before examining some of the particular forms which the theory assumes, it would be helpful to present a concrete example of the way in which a typical example of remembering would be dealt with. When a person knows, or is quite familiar with a particular piece of music, it is not at all unusual for the person to be able to call this piece of music to mind. He can, for example, hum the piece while driving his car. Or he can call it to mind in the sense of 'hearing it in his head'. Now this sort of ability is a power of memory. The person who is able to do this sort of thing is commonly said to remember the particular piece of music in question. Someone might, for example, ask me if I remember the opening bars of 'The Fourth Brandenburg Concerto' by Bach. And I might well reply, "Yes; it goes like this." I might then hum it for him. Now not everyone can do this. But many people can. And the ability to do this is not at all out of the ordinary. And likewise many people can 'hear in their heads' a piece of music which they know quite well. Not everyone can do this; but people who have some amount of musical ability can do this effortlessly. And this power is not considered to be unnatural. It is this sort of case which I wish to examine at first. It provides a very clear example of a typical trace theory explanation, since we do have machines which record traces of sound waves. (That is, we don't have machines which record and play back smells.) Just to avoid confusion, I mean to be considering a case where the person is so familiar with the music that he can call to mind a particular performance of the piece. He seems to hear in his mind the actual performance, clear in every detail. For example, if the piece is played by an orchestra, the person seems to hear the music as played by the very instruments used by the orchestra. For instance, when I call to mind a particular performance of 'The Fourth Brandenburg Concerto,' I hear the opening bars played on instruments called recorders, and not on some other instrument (trombones, for example).

The trace theory would give the following account of what happens when I 'hear in my head' 'The Fourth Brandenburg Concerto' (hereafter abbreviated as '4b'). To remember the opening of the '4b' in all its glory, to actually '*hear*' the recorders, the violins, is only possible if somehow there is an existing record of the '4b' within me. There *must* be a record, a trace which, by a certain way of speaking, carries with it all the 'information' necessary for reproducing the '4b' otherwise, how could we explain my ability to reproduce the '4b' so accurately? There must be an elaborate mechanism which records a trace (also called an 'engram').

But first, we will need a short account of the physics of the musical sound waves, for the trace theorist's explanation assumes this physical account as background.

The sound of the orchestra reaches my ears as a very complex and subtle intermix of various frequencies of sound waves, present in varying strengths. For example, if the instrument known as the recorder is playing a note of pitch *C* at 512 cycles per second, then it (the recorder) also produces other frequencies. The 512 cycle tone is called the 'fundamental', and the other subsidiary frequencies produced are called 'harmonics'. The recorder playing a pitch of *C* at 512 cycles will also produce, for example, harmonics at frequencies of 1024 and 2048 cycles per second. It might be, for example, that 96% of the sound energy is carried in the fundamental 512 cycle tone, while 2.6% of the energy is carried by the 1024 cycle harmonic, and .43% of the energy is carried by the 2048 cycle harmonic, and smaller amounts of energy are carried by various other harmonics. Suppose, on the other hand, that the violin were playing the same note of pitch *C* at 512 cycles per second. It would produce a different proportion of the various harmonics. (e.g., 90% at 512 cycles, 5% at 1024, .39% at 2048 cycles, and so on – The figures are, by the way, fictitious examples.) Indeed, a very important factor in the human ability to recognize various instruments playing the same note is this differing proportion of harmonics in the different instruments. This is not speculation, but is classical acoustics. Studies have been done in which (e.g.) tapes of various instruments have been played back without the harmonics (with the harmonics filtered out); and when the tapes are played back, people have great difficulty in identifying the various instruments being played. So far as I know, the importance of harmonics is undisputed among those who have studied acoustics. And while this account is simplified, it is complete enough for our purposes. Though a partisan of the trace theory would undoubtedly offer an account something like the preceding, it is not part of the theory itself. Rather, as I just said, it belongs to classical acoustics.

So how can it be that when I call to mind the '4b', I hear (in my 'inward ear') the recorders as recorders and the violins as violins, and not, e.g., as trumpets or pianos or saxophones? It *must* be that I am 'listening' to a trace, a memory recording, something very closely analogous to a tape recording of the concerto. Indeed, this trace must be very precise, and it must preserve all the subtle interplay of all the fundamentals and harmonics in their proper proportions. Now when the sound waves reach my ears, they cause small hairlike structures of appropriate lengths to vibrate with the different frequencies. These filaments are connected to nerve endings which fire when stimulated by the movement of the filaments. (Or perhaps the nerve endings fire all the time, but fire in a different manner when the filaments are moving – the details are unimportant.) There are filaments of varying length over a wide range of frequencies, including the frequencies of the fundamental and harmonic tones. So this structure, very like a microphone, is the means by which sound waves are transformed into electrochemical nerve impulses. These impulses are sent to the brain. If the auditory nerve which carries these impulses were to be severed, then, of course, deafness would result. Now it *must* be that somewhere in my brain (or central nervous system) these electrochemical impulses leave some sort of recording or trace. And later, when I remember the concerto, this trace is somehow activated, and it sends out duplicate nerve impulses. (i.e., nerve impulses nearly identical to the ones which which recorded the trace). Since these 'memory impulses' do *not* originate in the ear (in the organ which transforms sound waves to electrochemical impulses) naturally their effect is not absolutely identical to the effect of the original impulses. That is, I don't find myself deceived into thinking that I am actually hearing a performance of the concerto.

That then, is a suitably general illustration of the trace theory view. I find it very tempting. It comes to me without effort. I grew up believing that something like it *must* be the case. It seemed that new scientific breakthroughs buttressed the theory; a good example of such a breakthrough is the discovery of the chemical mechanisms of heredity – DNA and RNA. Surely memory, too, would turn out to be a mechanistic phenomenon. And, of course, this way of putting the matter makes it seem as if the only alternative view would be that memory is a nonmechanistic, perhaps supernatural phenomenon. One either opts for a mechanistic view or else one is forced to support some sort of vitalistic theory of disembodied pure spiritual substances. In other words, one has to make a choice between science and nonsense. I believe that this view of the matter is very widespread. In conversations with friends, I often find that they have the same reaction I did. Some sort of trace theory must be true. It has been said that theories concerning memory have been moved from the domain of philosophy to the domain of science. This is the view expressed in a book called *Consciousness and Behavior.*[3] Trace theory is so well embedded among other scientific theories, that Herbert Feigl finds it natural to use it as an example of an almost undisputed theory with no serious competitors:

> Customarily it is said, for example, that visible light is electromagnetic radiation, that table salt is NaCl, . . . that (at least) short range memory traces are reverberating circuits in cerebral cell assemblies, etc.[4]

So the assumption is that, however much philosophical speculation went on in the past (speculation about memory, and whether or not it might be a mechanical phenomenon), we are now dealing with a full-fledged scientific theory. And from Feigl's remarks we would presumably be justified in concluding that the theory was not only *bona fide*, legitimate science, but also that the theory had been tested and confirmed quite roundly. However, my own impression is that the experiments which have been performed, and which are cited as supporting the trace theory (e.g., Penfield's work, mentioned previously, or Lashley's which we will look at a little later on) certainly constitute no such thing as a confirmation, or even a support, for the theoretical structures heaped upon them. But the weaknesses of the theory run much deeper than a mere lack of decisive evidence.

It is this assumption (i.e., that trace theory is science, not philosophy) which I wish to question. Though trace theories present the *appearance* of being hard-nosed scientific theories, I hope to convince the reader that this is mere illusion. First, however, it would be a good idea to take a look at some of the different expressions of this sort of view found in current literature. One very common version of trace theory hypothesizes that the trace would be stored in the form of electrochemical changes at the junctions between nerve cells (this junction is called a 'synapse'). Culbertson (*op. cit.*, pp. 153–4) claims that 'variations at the synapse' could provide the basis for human memory. H. Chandler Elliott makes the more specific claim that the variations would take the form of altered electrical resistance at the synapse. The idea is that, after a memory trace has been laid down at the synapse (or synapses), then the next time a particular discharge runs through the synapse, the reaction of the circuit will be different from the way it first reacted. For example, suppose that the first time a child sees a fire, he reaches out to touch it and is burned. Now the next time he sees a fire, he will probably not reach out to touch it. Elliott's claim would then be that the incoming visual nerve impulse (when the child originally looks at the fire) causes electrical changes in the resistance of certain synapses. And the next time the child sees the fire, the incoming visual nerve impulses go over electrical pathways that are altered by the first impulses.[5]

Both Culbertson and Elliott opt for what we might call 'structural' trace theories, where the trace is supposed to be a more or less static structure (in this case a modification of the synapse). But there is another, perhaps more sophisticated form of the theory which maintains that the trace is not a thing with a relatively permanent and static structure, but rather a dynamic pattern of electrical discharge. The memory trace, according to this view, is actually a patterned, or 'modulated' current running constantly from cell to cell, or perhaps within one or more cells. This is probably what Feigl refers to in the paragraph quoted earlier. I do not see how, by the way, he can make this claim with such confidence. So far as I know, there is still an unresolved question among memory theorists, with regard to whether traces are dynamic or static entities. But I think it is unimportant for Feigl, who would probably admit that all he was claiming was that *some* form of trace theory is, beyond a reasonable doubt, correct. At any rate, both views are current in the literature. In the *Neurophysiological Basis of Mind,*[6] Sir John Eccles (whose stature as a neurophysiologist leads others in the field to quote him frequently) feels (or felt twenty years ago when the book was published) that a static rather than a dynamic theory was more likely to be correct. He felt that changes in the synaptic knobs of the nerve cells constituted a more plausible memory mechanism than 'self-maintaining electrical patterns'. K. S. Lashley, on the other hand, favored the idea of a dynamic trace, consisting of a 'pattern of electrical activity'. All brain cells are firing all the time, he says.[7] But the situation is apparently complicated by the fact that the synapses are not the only possible locus for structural changes. B. Delisle Burns[8] did not want to be more specific than to say,

> . . . the conclusion is unavoidable that the most probable pathways available to afferent nerve impulses within the Central Nervous System must have changed after something is learnt.

And W. R. Russell suspected that nerve fibers in the cortex might cross one another in such a manner as to act like synapses, thus greatly increasing the number of possible pathways for nerve impulses.[9]

Perhaps the most sophisticated form of the theory of static traces concerns the discovery of the mechanisms of heredity. Some neurophysiologists and neurobiologists have suggested that these same mechanicms (or similar ones) might also account for memory abilities. In a recent collection of essays we find two of the contributors suggesting this possibility. In the Introduction to this volume, K. H. Pribram gives very clear expression to the idea that memory theories and research are now bona fide science; so first we will look at the Introduction, and then at his remarks on the 'RNA memory trace theory':

The brain's unique power is in part due to its ability to store information, i.e., to store a coded representation of experience for future use. Not so many years ago Lashley (1950) reviewed the experiments which had been performed to show just how the brain goes about memory storage and retrieval. He came to the wryly stated conclusion that, on the basis of the evidence at hand, learning and remembering were obviously impossible. As can be seen from the sections in this volume, the situation has radically changed in these past two decades. True, we are still a long way from specifying the nature of engrams. However, now the techniques to investigate the problem have been developed, and the search is sophisticated and vigorous.[10]

And in an article in the same volume, Pribram says,

Biological scientists the world over have recently turned their heavy laboratory artillery on an age-old problem: the nature of memory mechanisms; event storage and retrieval; learning through novel experience . . .
The well-known role of the RNA molecule, together with its more stable sister substance, DNA, in the mechanisms of genetic 'memory' (his italics) stimulated the suggestion that RNA is somehow involved in the mechanisms of neural memory[11]

In the same volume, H. Hyden echoes these thoughts. He begins his article with what seems to me to be a clear and representative statement of the general outlook of trace theory. Accordingly (as was done in Pribram's case), I will first quote the beginning of his article, then the later part which deals with his specific suggestions.

The outer world is consciously experienced through the sensory part of the nervous system, and the information received is stored by memory mechanisms for future use.[12]

. . . We perceive the outer world and its stimuli, with the sensory part of the central nervous system, producing a 'change by use' (his italics) in a memory mechanism which stores the accumulating experience.[13]

The capacity to recall the past to consciousness can certainly be expected to reside in a primary mechanism of general biological validity. (At this point, Hyden goes into a discussion of the genetic mechanism of RNA production, and its plausibility as a memory mechanism.)[14]

In the past few years, the trend in memory research has been toward postulating the existence of more than one memory mechanism. It is well known that many people can memorize things for a relatively short time, but cannot retain the memorized items for very long. An example of this sort of phenomenon would be the power to retain a telephone number for a minute or two. After dialing the number, people are often not able to recall it. Observations of this sort have led theorists to postulate the existence of a short-term memory store – one which stores information in some relatively unstable and easily disruptable form. If the item is kept in this store for a long enough time, it may then be transformed into a relatively stable long-term memory.

First, newly presented information would appear to be transformed by the sensory system into its physiological representation . . . and this representation is stored briefly in a sensory information storage system. Following this sensory storage, the presented material is identified and encoded into a new format . . .[15]

In modern terminology, there must be one mechanism which receives and loses impressions with great rapidity, as the retina does, and another which retains impressions throughout the greater part of a lifetime.[16]

First, we need to distinguish between short and long term traces . . .[17]

Since the postulated short-term trace seems to be so easily disturbed or destroyed, some theorists assume that the trace (the short-term trace) must be of the 'dynamic' type – some sort of electrical discharge which circulates for a small time before either being lost or copied into the long-term trace. The more stable long-term trace is naturally assumed to be a structural entity. So this leads to a combination of the structural and the dynamic trace theories.

This modulated carrier would continue to circulate through networks for some time, functioning as a short-term storage mechanism and achieving consolidation by repeated passages of the information over the elements implicated in the formation of the permanent engram.[18]

In general, however, the longer an item remains in primary memory, the likelier it is to have been copied into secondary memory . . . The traces of items stored in secondary memory do not seem to decay in time . . .[19]

The bilateral medial temporal lobe lesion appears to disturb selectively an essential transition process, or process of consolidation, by which some of the evanescent information in primary memory acquires an enduring representation in the brain.[20]

Various data seem to require further, that the postulated coupling between electrical patterns and the long-term storage device be reversible – that the pattern of iterated or sustained electrical activity stipulate some representational structural modification, and that this structural modification be able to generate an electrical pattern identical to the one which established it.[21]

The preceding passages are concerned with a short-term trace which has a lifespan of seconds or, at the longest, several minutes. But, as we all know, may people can remember long lists of items, keeping this knowledge for about a day. And if the knowledge was memorized for some specific purpose (imagine a student memorizing a list of twenty authors' names for a spot-quote test), and if it is not called up after the one use, it is often forgotten in a day or two. This phenomenon is familiar to students. Now it might lead us to wonder about the distinction between short and long-term memory. We might, for example, be of the opinion that people remember all sorts of different things for all sorts of time periods. For example, a young man who is pressing his attentions upon a young lady might be quite familiar with her telephone number, since he dials it ten or twenty times a day. But a few weeks after the end of the affair, he will quite likely be unable to remember it. (The phone number, that is.) Are we supposed to immediately assume that there must be a separate trace system whose traces are no longer activated? It seems plausible to suppose that this method of postulating trace mechanisms would quickly lead us to invent (or discover) a large number of them. This doesn't seem to trouble some trace theorists, however. As they realize that memory phenomena are more complex, they merely add parts to the postulated mechanism.

This fact suggests to me that perhaps we will have to entertain the possibility of still another storage system, the repository of traces so well entrenched that they can be revived automatically without an intervening process of verbal search.[22]

Others feel that the distinction between short and long-term memory is too crude, and there is, in fact, an intermediate-term memory between the other two, making a total of four memory systems in all. . . . Whether there be two memory systems or four, everyone does agree that newly presented material is forgotten rather rapidly unless it is rehearsed.[23]

Wayne Wickelgren . . . proposes a generalized version of a strength theory of memory with four different types of memories: a very-short-term, a short-term, an intermediate-term and a long-term memory.[24]

Wickelgren's proposal is the most elaborate version of trace theory I have seen.

Each trace . . . passes through four phases . . . The acquisition phase refers to the period of presentation or active rehearsal of events during which the memory traces are initiated. Some consideration is given to the nature of the coding for events and associations in different modalities by making provision for similarity functions between pairs of events and pairs of associations. As an example of event similarity, the letter names 'B' and 'D' are more similar in phonetic short-term memory than 'B' and 'S'.[25]
Every sensory, motor and cognitive modality of performance is a modality of memory. In each modality of memory there are as many as four traces with different time courses: very-short-term, short-term, intermediate and long-term memory.[26]

There are, however, several competing theories which claim to explain the different 'lifetimes' of different memories. Some of these theories make do without postulating the existence of many memory storage systems.

> There are several reasonable ways in which rehearsal might act to increase the resistance of a memory trace to interference; . . . Rehearsal might increase the strength of a single memory trace, it might increase the number of traces of the item, or it might accomplish a change from short-term storage to long-term storage in the state of the item.[27]

Harley Bernbach of Cornell presents us with an example of a theory which postulates only one memory store. In his theory, an item which is remembered for a long time (an item which 'resists fading') is one which has many copies in the memory store. And the items which have the most copies are those to which the person paid most attention when they (the remembered items) were presented. We can imagine someone suggesting that, if Bernbach's theory is correct, science would at last be provided with an objective and quantitative measure of attention. The more memory-trace-replicas present in Mr. A's brain, the more attention he paid to that item when he first encountered it. At any rate, Bernbach begins his article this way:

> This chapter presents a model for postperceptual verbal memory that postulates a single memory store, with multiple copies, called *replicas* (his italics) created in memory by rehearsal processes.[28]

Bernbach is not the first to propose that there might be more than one copy of each trace. Lashley, for example, felt that the trace must be represented throughout entire regions of the brain. He came to this conclusion after thirty years' fruitless search for an isolated trace. He found that a rat (e.g.) would still react correctly to a cue (e.g., a green triangle) it had been trained with, even though large portions of the relevant brain structures had been removed. At the point when the rat no longer reacted to the triangle (say, when 95% of the visual cortex had been removed) the rat no longer reacted to *any* kind of visual stimulus. So Lashley concluded that his removal of the brain tissue was not the removal of discrete memory traces.[29]

We have seen how some trace theorists have invented more complex theories, with various different postulated structures and mechanisms for storing memory traces. There are also theories which postulate something other than the simple storage (as in a tape recorder) of recorded traces. It has been suggested, for example, that engrams are not only stored, but combined (whatever that means) in various ways, as a part of normal brain activity.

In this view Brickner concurs after an excellent and thorough analysis of the behavioral defects in man following bilateral frontal lobectomy, and indicates that 'all the interpretable changes may be explained by a diminution in the associative function of synthesizing simple mental engrams into more complex ones'.[30]

There are also variations of trace theory in which it is not only engrams that are stored in the brain, but other, somewhat surprising things, as well:

There are, of course, many complexities here, both of engram formation and motivation. But the point is clear: the effects produced by stimulation of the higher central nervous system are subject to lasting alteration by prior, associated activity. Furthermore, by such regulated intrusions into the normal activity of the brain, it is possible to manipulate both engram and motive to ultimately reveal something of their neural nature.[31]

I propose that the first stage of memorizing an association involves storing a representation of the stimulus-response pair in memory as a unit.[32]

Both theories referred to in the preceding quotes are combinations of stimulus-response and trace-type views.

This concludes a short introduction to trace theory, and a small survey of some current versions of the theory. In the next chapter, the theory will be further characterized. The chapter concludes with a criticism of trace theory.

NOTES

[1] Thomas A. Harris, *I'm O.K. – You're O.K.* (Harper & Row, 1967, 280 pp.).
[2] Wolfgang Köhler, *Gestalt Psychology* (Liveright Publishing Co., 1947).
[3] J. T. Culbertson, *Consciousness and Behavior*, W. C. Brown, Dubuque, Iowa, 1950).
[4] Herbert Feigl, 'The mental and the physical', *Minnesota Studies in the Philosophy of Science*, Vol. II, – *Concepts, Theories and the Mind-Body Problem*, ed. by H. Feigl, Michael Scrivner and Grover Maxwell, University of Minnesota Press, Minneapolis, 1958, 553 pp., p. 438.
[5] Elliott discusses this on pp. 33 and 39 of his book, *The Shape of Intelligence – The Evolution of the human brain*, Charles Scribner's Sons, New York 1969).
[6] J. Eccles, *Neurophysiological Basis of Mind* (Oxford, 1952, p. 226).
[7] He discussed this in an article entitled 'In search of the Engram', See pages 502–3 of a book edited by F. A. Beach and D. O. Hebb. The book is entitled: *The Neuropsychology of Lashley*.
[8] B. Delisle Burns *The Mammalian Cerebral Cortex*, 1958, pp. 79–80.
[9] W. R. Russell, *Brain Memory Learning*, Oxford Press, 1959, p. 17.
[10] K. H. Pribham, p. 7 of the Introduction to *Brain and Behavior 3: Memory Mechanisms*, ed. K. H. Pribram, 1969, Penguin Books, 524 pp.
[11] K. H. Pribram, 'The new neurology, memory, novelty, thought and choice', in *Brain and Behavior, op. cit*., pp. 54–55.
[12] H. Hyden, 'Biochemical aspects of brain activity', in *Brain and Behavior, op. cit*., p. 33.
[13] p. 45, *op. cit.*
[14] p. 45, *op. cit.*
[15] *Models of Human Memory*, ed. Donald A. Norman, 1970, University of California, San Diego, Academic Press, 537 pp. From an article by Donald A. Norman, entitled 'Models of human memory', p. 2.
[16] *The Science of Mind and Brain*, by J. S. Wilkie, 1953, Hutchinson's University Library, London, 160 pp., p. 42.
[17] *Models of Human Memory, op. cit*., article by Donald A. Norman and David E. Rumelhart, 'A system for perception and memory', p. 21.
[18] *The Biology of Memory*, ed. by Karl H. Pribram and Donald E. Broadbent, 1970, Academic Press, London and New York, 323 pp. from an article by M. Verzeano, M. Laufer, Phyllis Spear and Sharon McDonald, 'The activity of neural networks in the thalamus of the monkey' p. 239.
[19] *The Biology of Memory, op. cit*., from an article by Nancy C. Waugh, of Harvard Medical School, 'Primary and secondary memory in short-term retention', pp. 63–4.
[20] *The Biology of Memory, op. cit*., from an article by Brenda Milner, Montreal Neurological Institute, p. 47.

[21] *Information Storage and Neural Control*, ed. by William S. Fields and Walter Abbott, publ. Charles C. Thomas, Springfield, Illinois, 1963, from an article by E. Roy John, 'Neural mechanisms of decision making', p. 244.
[22] *The Biology of Memory, op. cit.*, article by Nancy Waugh, 'Primary and secondary memory in short-term retention', *op. cit.*, p. 65.
[23] *Models of Human Memory, op. cit.*, article by Donald A. Norman, *op. cit.*, p. 2.
[24] *Models of Human Memory, op. cit.*, article by Donald A. Norman, *op. cit.*, pp. 10–11.
[25] *Models of Human Memory, op. cit.*, article by Wayne A. Wickelgren, 'Multitrace strength theory', p. 66.
[26] W. Wickelgren, *op. cit.*, p. 70.
[27] *Models of Human Memory, op. cit.*, article by Robert A. Bjork, 'Repetition and rehearsal mechanisms', p. 324.
[28] *Models of Human Memory, op. cit.*, article by Harley A. Bernbach, 'A multiple-copy model for postperceptual memory', p. 102.
[29] See *The Neuropsychology of Lashley, op. cit.*, the article, 'In search of the Engram', especially p. 501.
[30] *Brain and Behavior, op. cit.*, article by C. F. Jacobsen, 'The functions of the frontal association areas in monkeys', p. 300.
[31] *Brain and Behavior, op. cit.*, article by R. W. Doty, 'Conditioned reflexes formed and evoked by brain stimulation', p. 183.
[32] *Models of Human Memory, op. cit.*, article by James G. Greeno, 'How associations are memorized', p. 259.

CHAPTER TWO

TRACE THEORY CRITICIZED

In the last section we presented an example of a typical trace theory-type explanation of how it is that one can remember (in the sense of 'hearing in one's head') 'The Fourth Brandenburg Concerto' (abbreviated '4b'). It should be clear by now that an essential part of trace theory's attractiveness is that it seems to offer a hard-nosed scientific account of what goes on in memory. Each step in the process is completely determined by what went before – a classically causal explanation. The steps in the process referred to are as follows: The orchestra plays various notes on various instruments. This, of course, causes patterns of disturbances in the surrounding air. If an instrument is (e.g.) struck forcefully, it gives off a correspondingly large disturbance to the surrounding air. This in turn would produce a correspondingly large displacement of the ear drum, resulting in a correspondingly large movement of the tiny filaments in the organ of the inner ear (which we have discussed already). Now it must be assumed that this relatively large movement of the filament causes a relatively large shift in whatever sort of electrical currents run through the auditory nerve. And a softer sound would presumably cause a relatively smaller disturbance. Any way in which two sounds differ must, it is assumed, be reflected in the nerve impulses aroused by them. Otherwise, the two sounds would strike us as the same, when in fact they are different. This step-by-step process must continue right on into the brain. Otherwise, if two different nerve impulses (differing in voltage, for example) caused exactly the same electrochemical brain reaction, the two sounds which caused these currents would necessarily sound the same. Now the memory trace laid down when I heard the '4b' must also obey these rules.

That is, the trace must differ from other memory traces (e.g., from traces of 'The Fifth Brandenburg Concerto'). Otherwise, when the trace of '4b' was activated, it would not cause a different brain current (or whatever) than would the trace of 'The Fifth Brandenburg'. And then, of course, the two would sound the same to me in memory. In fact, the trace *must* be of such complexity that all the subtle shifts in harmonics are captured. Otherwise, I might remember the '4b' as being played on an electric guitar. And if the music grows loud at a certain point, then there must be a corresponding feature incorporated into the trace. Otherwise, e.g., when people sing 'The Star-Spangled Banner', they might sing a syllable *very* loud every once in a while. If you pause to consider, this sort of thing definitely does not happen in a normal case. (Imagine someone singing, 'Oh say CAN you see . . .'). The trace theorist assumes that this fact (i.e., the fact that a normal person would *not* sing the 'can' much louder than the other words) is the result of a straightforward causal connection between the original sound, the trace, and finally, the singing of the remembered song. Put this way, it seems very convincing. How could anyone ever remember anything correctly, unless the brain had stored the information to be remembered? In the case of calling to mind the '4b', how could it sound like a baroque orchestra unless the trace had some feature corresponding to (e.g.) the recorders, the violins, etc.? Perhaps the most forceful way to put it is this: It is not possible that the brain could produce spontaneously, out of the blue, just *those* electrochemical patterns corresponding to a baroque chamber orchestra performing the '4b'! Why, to suppose that such a thing could happen would be just like supposing that we could put a blank tape on a tape recorder, set the machine in motion, and find that the machine was producing, spontaneously and at random, just *those* electrical patterns which correspond to the notes of the '4b'. And that sort of supposition is nonsense, mumbo-jumbo.

The requirement that the trace have a structural feature for each feature of the remembered piece of music, is the requirement that the trace be *structurally isomorphic* to the piece of music. Wolfgang Köhler is very explicit in his discussions of the idea of isomorphism.

Obviously, however, if the characteristics of concomitant physiological processes are to be inferred from given characteristics of experience, we need a leading principle which governs the transition. Many years ago, a certain principle of this kind was introduced by E. Hering. Its content is as follows. Experiences can be systematically ordered, if their various kinds and nuances are put together according to their similarities . . . The processes upon which experiences depend are not directly known. But if they were known, they could also be ordered according to their similarities.[1]
In 1920, the Gestalt psychologists transformed this assumption into the following general hypothesis. Psychological facts and the underlying events in the brain resemble each other in all their structural characteristics. Today we call this the hypothesis of Psychophysical Isomorphism.[2]

This principle is not restricted in application to the field of memory, but is meant to apply generally to all psychological phenomena. Köhler offers instances of the way in which the principle is to be applied. On page 37 of *Gestalt Psychology* (*op. cit*) for example, he says that experienced loudness must vary with some aspect of some brain process. And on page 38 he says that, whatever brain process corresponds to color perception, it must be variable in as many dimensions as color-experiences are. Otherwise there would be a lack of isomorphism between experience and brain process. Why would a lack of isomorphism be so unacceptable?

Suppose someone looks at a red paint spot, and then at a green spot. Now one absolutely fundamental reason that the two spots are perceived as different in color is that photons of different wavelengths strike the retina of the observer's eyes. It is assumed that green light striking the retina affects the retina in a different way than red light does. Otherwise, the two would appear to us as being the same color. Now, is it plausible to assume that things are any different in the brain? In other words, not only must there be a difference in the *retinal* processes set off by red and green light, but there must also be a difference in the nerve impulses relayed to the brain, and then in the brain processes themselves. If red and green light caused the retina to send the same impulses to the brain, then how would the brain (the visual center) be aware that the light is one color rather than the other? That would be just like building a car with a transmission such that reverse and first gear were the same. So when you push the 'reverse' button, the exact same lever is moved as when you push the 'first gear' button. No mechanic in his right mind would expect this car to act differently when one button is pushed rather than the other. After all, both buttons move the very same lever (in exactly the same way, of course). Just labeling one button 'reverse', and one 'first gear' doesn't make the car go in one direction rather than the other. The two buttons must cause two different movements in the car's transmission. It would be a violation of very basic notions of cause and effect if the car's designer told us, "Well, even though the two buttons are connected so that they do the same thing, nevertheless the car goes forward in the one case, and backward in the other. Strange, isn't it?"

So if the retina is struck by green, rather than red light, there must be some indication of this fact in the impulses that travel down the optic nerve; and this feature of the nerve impulses must also cause some corresponding difference in the brain process which it sets off. This would, of course, be true not only of color, but of all other 'aspects of experience'.

> Experienced order in space is always structurally identical with a functional order in the distribution of underlying brain processes. This is the principle of *Psychophysical Isomorphism* in the particular form which it assumes in the case of spatial order (his italics).
> Experienced order in time is always structurally identical with a functional order in the sequence of correlated brain processes.[3]

For Köhler, it is not only brain processes and subjective experience which exhibit this ismorphism. Rather, his concept is that of a chain of isomorphism, stretching from the physical (outside) world to the brain processes, and then to the world of experience. Consider his account of what goes on at a piano recital:

> Where the pianist ends a phrase and starts a new one, he gives the sound waves such relations of temporal proximity, intensity, and so forth, as are likely to establish the same articulation in the auditory fields of people in the concert hall.[4]
>
> Hence, innervation projects upon the pianist's muscles an organization which his mental processes and their brain processes have in common. In this fashion, the formal relations among the resulting sound waves are determined. But auditory organization in the people who listen depends upon such relations. Consequently, their experiences tend to be organized in a way which agrees with the organization of mental processes in the pianist.[5]

We can imagine an unbroken chain of isomorphism extending from the mind of the composer, to the music printed on paper, to the mental processes of the pianist, to the sound waves, and finally, to the brain processes, and mental processes, of the listener. The trace is the bearer of this isomorphic structure when the music is not being consciously remembered. And when the music is called to mind, it is assumed that the trace is somehow activated to give rise to brain processes isomorphic to the ones which were caused by the sound waves when the music was actually heard. To deny the existence of an isomorphic trace is to say that the brain could spontaneously produce processes (impulses, currents or whatever) isomorphic to the ones produced by the sound waves. And this would be tantamount to saying that the causal chain could be incomplete. So the requirement that there be a trace is the requirement that the record player must have a record placed upon it before it can play music.

By this time the reader will have a pretty fair notion of what trace theory is, and also of just how widespread is its support. Now the question will be asked, "How can someone who is not a neurophysiologist (or some sort of scientist who has studied the structure of the human nervous system) possibly criticize a neurophysiological theory of memory?" Isn't this a bit like some illiterate person trying to tell a great poet how to write a poem? Or perhaps a clearer and more relevant example is offered in the case of a cheeky person who never studied chemistry beyond high school telling a research chemist that he is going about his research in entirely the wrong way. As I said before, I don't believe that trace theory really is what it seems to be. I think that we will see that it turns out to be philosophy, not science. And further, it turns out to be bad philosophy, the result of misconceptions, illusions, and unclear thinking. Yet it is obviously a natural path to turn toward in an effort to understand memory. And it certainly *looks* scientific. It seems as if the question (as to whether or not trace theory is correct) were entirely empirical. That is, someday scientists will have done enough brain research, and will be able to open a skull and say, "Look, philosopher. Here is the recording device." And I certainly cannot claim to have looked into any brains. So why do I think that trace theories do not provide an explanation of human powers of memory?

The first step in the attack is to follow through on one of these examples of a trace-theoretical case of memory, and to see just where it leads. This will help to get rid of the idea that there *must* be a trace, that the theory *must* be correct. Accordingly, let us return to the example (presented on pages 5–8) of a trace-theoretical explanation of a particular case of remembering. This involved calling to mind a piece of music, in the sense of 'hearing it in one's head'. The example involves the assumptions of classical acoustics – assumptions concerning the nature of the sound waves, fundamental and harmonic tones, etc. The principle of 'psychophysical isomorphism' is assumed to hold, too. That is, the trace must bear with it all the information necessary to reproduce 'The Fourth Brandenburg Concerto', in a full-blown baroque orchestral rendition. If the trace lacked this information (if it were incomplete in any aspect of its isomorphism) then we could not explain how it is that I am able to call to mind just that; i.e., a baroque chamber orchestra playing these notes, on just these instruments, getting loud here, softer there, etc. And we can see that the trace is only an intermediate step in this process. It is supposed to function as follows: the trace is activated to produce brain patterns (nerve impulses, or whatever) isomorphic to the ones which originally laid down the trace. So the principle of isomorphism extends beyond the trace on both sides. That is, when I originally hear the '4b', my brain processes must be isomorphic to the music as I hear it. And when I am recalling the piece (when I am 'hearing it in my head') it is assumed that, once again, there must be nerve impulses isomorphic to the music going on in my brain. Any time I am hearing the music (either in memory or 'in the flesh') there must be some brain process going on which mirrors the music, right down to the last detail. And such a brain process, complex and highly ordered, could not possibly spring into existence spontaneously; it must be the result of some causal interaction. In the case of *actually hearing* the '4b', the brain pattern (process, etc.) results from the nerve impulses which travel from the ear to the brain. In the case of *remembering* the '4b', the brain process must be the result of activation of the trace. That is the trace theory. Somewhere in my brain there must be a detailed recording of the '4b'; otherwise, how could we explain the ability to call to mind a detailed rendition of it?

But if we consider for a moment, I can also call to mind this very same musical passage, but with the lead part being played by trumpets instead of recorders. I maintain that I have this power, and I maintain further that it is no extraordinary power. Many people can do the same, with any piece of music they are familiar with. This power is not especially remarkable. Now, to call this creation to mind (i.e., the '4b' played with trumpets rather than recorders as the lead instrument) and to have it sound like *trumpets* playing the lead, then the impulses, wherever they are in my brain, *must* (on this view) be isomorphic to the sound waves of an orchestra-and-trumpets rendition of the '4b'. But how could these impulses (this brain process, pattern, etc.) possibly be produced? Is there a model, a trace, for these impulses? We are assuming that there is no *memory* trace of such an event, for I have never heard the '4b' played with a trumpet doing the lead part. And the trace theory holds that, in order to hear (in my mind) the '4b' played with trumpets, there must be impulses isomorphic to those produced by the actual hearing of such a performance. And in order to be able to produce those impulses, there must be a brain trace isomorphic to them, which serves as a model. And the trace of '4b' played with trumpets (let us call it '*4b+*') is different from the trace of '4b' played with recorders. This is so because a trumpet playing a note (a fundamental tone) will produce different harmonics than a recorder playing the same note. That is why a trumpet sounds like a trumpet, and not like a recorder. Now no one could maintain that I have a trace of '4b+' (since I have never heard '4b+'). So it must be that my brain (according to the assumed principle of psychophysical isomorphism) somehow creates, *without a compete model*, brain impulses isomorphic to '4b+'. How could this be?

Well, a defender of trace theory could say that the impulses isomorphic to '4b+' (which are produced by the brain when we call to mind '4b+') are produced in a slightly different way than in the case where what is called to mind is something that was actually heard in the past. In the case of calling to mind the '4b' (the piece of music I am familiar with) the detailed trace is simply activated to produce impulses isomorphic to it. But in the case of calling to mind '4b+' (the familiar piece of music, but played with trumpet instead of recorder), the trace of '4b' is indeed activated, but in a slightly different way. It does not in this case produce impulses *completely* isomorphic to itself. Rather, the impulses produced would differ from the impulses of '4b' in that they would contain elements corresponding to the differing harmonics of trumpet as opposed to recorder. Now how could this come about?

The defender of trace theory could say that, after all, there must be many memory traces of trumpets in my brain, and of course there is the trace of '4b'. So, by some fairly complex process, my brain combines these trumpet traces with the trace of '4b', and merely substitutes into the trace of '4b' the appropriate harmonic blends for trumpet instead of recorder. Now this move by the defender seems unobjectionable, for, after all, the brain is a marvelously complex thing, isn't it? All it needs to create the impulses isomorphic to '4b+' are the traces of '4b' and various trumpet traces.

But I want to point out that something very important just happened. A radical step was taken here which the defender of trace theory might not have realized. Impulses isomorphic to '4b+' were created by the brain, without there being a trace isomorphic to '4b+'. Specifically, there was no trace present which both had the harmonic blends of a trumpet, and which was in all other respects isomorphic to '4b'. Now, if the trace theorist admits that this is possible in the case of calling to mind '4b+', what does this show about the case where I call to mind '4b'? That is, the trace theorist must admit that '4b+' can be called to mind with only (1) a model which lacked the proper information about harmonic blends, and (2) various trumpet traces not all all similar to '4b+' except in that the harmonic blends are those of typical trumpet music. Now why can't this same sort of process take place in the case where I call to mind (remember) '4b'? Certainly there is no reason for the trace theorist to say that such a thing does not happen. In other words, if the brain can produce impulses isomorphic to '4b+' without a complete trace, then the same sort of incomplete trace would suffice to produce impulses isomorphic to '4b'. To call to mind '4b', I might only have (1) a trace which lacked harmonic elements, but which was otherwise isomorphic to '4b', and (2) various traces of typical recorder music. The *trace* need not have features corresponding to the harmonic overtones of a recorder; that could be added on. A Moog synthesizer is capable of just such a feat. A melody can be played with various mixtures of overtones, which makes it sound as if various instruments were playing that melody. In fact, this is what happens when the stops

are pulled on an organ. It should be obvious that this possibility applies to all of the instruments in the orchestra. That is, the trace of '4b' might contain no information about the harmonic overtones at all. It might just have elements corresponding to the fundamental tones which each instrument plays. Then the only other ingredient needed would be various unrelated traces of the various instruments in the orchestra. It would do, for example, to have traces of a violin playing scales, and a recorder playing a Beatles song. The reasoning presented on the preceding page constitutes an important step, so perhaps we should go through it again, just to avoid confusion. The defender of trace theory would have to admit that I need not have a trace isomorphic to '4b+' in order to be able to hear '4b+' in my mind. He has to admit this because first, I *can* call to mind '4b+', and second, I have never actually heard '4b+'. And if the machine (the brain) can do this in the case of '4b+', then it can do this in the case of '4b', too. So the defender of trace theory must admit the possibility that when the original trace of '4b' was laid down, it lacked information with regard to the harmonic overtones produced by the various instruments. My brain might have supplied *that* part of the music from other traces of recorder music, violin music, etc.

Before carrying this examination any further, a possible source of confusion should be eliminated. We can imagine a trace theory which postulated that the trace be separated into various parts. What that might mean is something like this: (e.g.) The trace is laid down with no information corresponding to the harmonics, *per se*, but with just a little indicator which would cause the 'recorder function' to add appropriate recorder harmonics to the melody represented in the trace. That kind of trace theory would be possible. And I am not sure whether or not a trace of this nature would be called isomorphic. But that is not the sort of point being made. My point against the trace theorist is much more radical than that. Surely there is no trace of '4b+' in my brain. (That is, there is no trace isomorphic to 'The Fourth Brandenburg

Concerto', with trumpet substituted for recorder.) I have never heard any such piece, so why would there be a trace of it? Whether or not memory traces have actual harmonics in them, or whether or not they just have little 'indicators' which determine which 'harmonic function' is to be applied, surely I do not have a trace of '4b+'. Yet I can call to mind '4b+' at will. So somehow my brain can, at any time, easily call forth impulses isomorphic to '4b', but with trumpet-type harmonics. So this same brain could presumably call forth impulses isomorphic to '4b', and with recorder-type harmonics, just as when I heard it. The trace of '4b' need have *no* indication of harmonic overtones, and my brain could still add on the appropriate (recorder) harmonics. After all, that is just what the trace theorist says it must do in the case where I call to mind '4b+'.

So the memory trace of '4b' *need* not be isomorphic in this respect (i.e., with respect to harmonics) either to the impulses which produced it when I actually heard the '4b', or to the impulses which are produced when the trace is activated. (I hope it is realized at this point that I am arguing from within the viewpoint of trace theory. I do not agree with the conclusions reached in these paragraphs, concerning the structure of the trace. These conclusions are steps toward throwing out the idea of a trace.) At any rate, the defender is forced to admit that the trace of '4b' might not contain information as to the various harmonic blends of the different instruments. The trace might, for example, be something like a representation of '4b', played on a tone generator (a machine which produces only fundamental tones, and no harmonics). And the defender has to admit that such an incomplete trace, not strictly isomorphic to '4b', could nevertheless be activated to produce impulses isomorphic to those produced when '4b' was actually heard. We can best characterize what has been done here in the following way. By the very assumptions of trace theory (especially the principle of psychophysical isomorphism), it has been shown that the memory trace might be more abstract than at first seemed possible. (At first it seemed as if the trace *must* contain *all* the information, or structural complexity, of the piece of music.) This process of abstraction can be continued by following through in our examination of the typical trace theory-type explanation.

Suppose, then, that we are correct, and suppose that the trace need not contain structural features corresponding to the harmonic overtones of '4b'. What is left? What else must the trace contain? When the music is first learned (when the trace is first laid down), the '4b' is played at varying levels of sound – louder here, softer here, and so on. And when I call it to mind, I hear it pretty much like the original. That is, I do not hear it with the opening bars played disastrously louder than the rest of the concerto. A further reflection of this fact is that, if I were to hum the '4b', I would not hum the first few bars disastrously loud unless I were trying to annoy someone, or something of the sort. So the trace theorist would say that the trace obviously contains information (structural features, etc.) which correspond to the loudness of the music. Since this loudness was a feature of the original music (and of the brain impulses produced by the sound waves) and since this feature of loudness is once again present when '4b' is called to mind, it must have been preserved in the interval. That is, the trace must bear some indication of relative loudness and softness of the music. Otherwise, how could anyone get it right when they call the music to mind, or hum it?

But it is very easy to call to mind a performance of the '4b' which is *much* louder than any I have ever heard. Or, alternatively, it is just as easy to call it to mind in a whisper-quite performance – one much quieter than any orchestra would ever play it. And this can be varied at will. I can call to mind the first note very loud, and the second note very soft. Here we have a rendition of '4b' with loud and soft passages different from the original. And so, by the principle of psychophysical isomorphism, the brain has created impulses which differ (in respect to the feature corresponding to loudness) from the 'loudness' feature of the trace. Once again the trace theorist will want to say that this offers no problem. Of course a complex record-and-playback system like the human brain would have a variable loudness control. But once again, a step has been taken which is more radical than it first seems to be. For the trace theorist has just admitted that the brain can, with no trouble, produce impulses isomorphic to '4b', except having a different distribution of loudness than the original. If so, then the brain could just as well produce the right pattern of loudness from a trace of '4b' which, for example, had some different distribution of loudness indicators. For example, it could produce an accurate '4b' (accurate with respect to loudness) from a trace of the '4b' which had each note as loud as the next. And that is to say, the trace need contain *no indication of loudness at all*. This point is precisely the same as the point about harmonic overtones. If the brain can create a '4b' with a *new* pattern of loud-and-soft, then the same process could be used in the case of calling to mind '4b' with the *usual* pattern of loud-and-soft. So the trace again might be more abstract than it seemed.

The reader will now perhaps realize that the process of 'whittling down' the trace does not end here. So far, two vital acoustic features have been shown to be *unnecessary* for the ability to call to mind the '4b' (i.e., for the brain's presumed ability to create impulses isomorphic to the '4b'). With the harmonic overtones and the loudness factor gone, what is left of the trace? Well, how about the melody? Surely the brain cannot produce some highly ordered and complex melody without a model to guide the tone generating part of the mechanism. This melody consists of a number of musical notes, spaced relatively closer or further apart in the appropriate scale. But this relative spacing between the notes can be changed. I can call to mind the '4b' with, for example, the last note of each phrase moved up an octave. Or I can call to mind '4b' with the last note of each phrase being replaced by some really awful sounding dissonance. (Think, e.g., of the orchestra stopping at the end of each phrase, allowing a trained seal to honk a horn on some dissonant note.) When I call one of these creations to mind, the trace theorist must suppose that my brain is producing a '4b' with a different relative spacing of the melody notes, different, that is, from the relative spacing recorded in the trace. It is important to see that this is not done at random. The new melody (even with the dissonant note) is just as complex and just as regular as the original melody of '4b'. And I can reproduce it at will. And though not everyone can do this sort of thing, many people can. And those who can are not looked upon with awe, fear or wonder. This is part of what is usually called 'having a good ear'. The upshot of all this is that the brain can (on the trace theorist's view) produce a complex and highly ordered melody without an isomorphic trace. This is what it does when I call to mind '4b' with the new melody. So why can't the brain do this when I call to mind '4b' with the usual melody? So the relative spacing of the melody notes need not be recorded in the trace. It would suffice, for example, if the trace had only an indication of how many melody notes there were; the relative spacing could be added on later, just as it is when I call to mind the dissonant '4b'.

This conclusion also applies to the absolute values (pitch) of the melody notes. That is, I can call to mind (or hum) the '4b' in various different keys. So the absolute value of the melody notes (expressed in cycles per second, if you like) need not be represented in the trace, either.

A few sentences back, I said that it would suffice if the trace contained simple markers to indicate how many notes there were in the melody. But really, even this feature is not necessary. I can call to mind '4b' with two notes instead of one at the end of each phrase, or with extra notes here and there. This would not be done at random; these extra notes would be repeatable, every time I felt like 'calling up' this new, embellished version of '4b'. In this case, no indication of these extra notes would appear in the trace. Yet, by the principle of psychophysical isomorphism, my brain produces impulses which correspond to these new notes in the embellished version of '4b'. If my brain does this in the case where I call to mind an *embellished* '4b', why can't it do the same thing when I call to mind the normal, *unembellished* '4b'? If the brain can add some particular note (actually an impulse corresponding to a note) to the trace of '4b', then it can add any note. For example, if my trace of the '4b' lacks some particular note, then the mechanism just adds it on. So my trace of '4b' may actually lack some melody note, which gets added on by the same part of the mechanism which adds notes on in other cases. But, of course, if the machine can somehow do this for one note, then why not for more than one? Maybe the trace of '4b' only has indicators for every other note, or every third note, or every fourth note, or maybe only for the first note. At this point, we can see that the trace no longer does any work in the explanation of how the brain produces impulses isomorphic to '4b'. The trace *need* not bear any indication of musical pitch, harmonics, loudness, or even how many notes the melody contains. The only other feature which the trace might need is some sort of indication of the time between notes, and the rhythm of the music. But surely it is no problem to call to mind a very fast '4b', much faster than I have ever heard it. The same is true for a slow '4b'. And it is just as easy to call to mind a '4b' with the rhythm changed (syncopated, for example). In any of these cases, the brain is assumed to be working some particular transformation upon the rhythm and/or speed of the original trace. But then, maybe the same process occurs when the normal '4b' is called to mind. Maybe the trace has some representation of rhythm or speed quite different from the usual way the '4b' is played. And the brain somehow works some particular transformation upon this unheard-of rhythm and/or speed to transform it into the '4b' which has the normal rhythm and/or speed. For example, the trace may contain a 'speed indicator' of a very high value. And it may have a rhumba rhythm. But when I call to mind the '4b', the speed-transforming mechanism and the rhythm-transforming mechanism go to work (just as they do in other cases of changing the rhythm and speed) and produce the normal '4b'.

So there is nothing left which the trace *must* contain. And that is just another way of saying that there need be no trace; for a trace which contains no 'information' (no structural features corresponding to features of the music) is no trace at all. So the brain is capable of creating impulses isomorphic to a piece of music without any trace of that music. This happens when, for example, I call to mind the '4b' played with trumpets too fast in a rhumba rhythm, very loud, with a couple of trills thrown in for embellishment, and perhaps the first note of each phrase an octave higher than it should be.

It would be useful to assess what has been done so far. We examined a typical example of memory (calling to mind a familiar piece of music) from the point of view of trace theory. The examination was general, in that we were not interested in any particular type of trace theory. That is, any observations made would apply equally well to static or dynamic trace theories, and equally to theories which postulate one storage system or more than one. An attempt was made to present trace theory in its most tempting aspect. The feature of trace theory which makes it so tempting is that it looks scientific. More specifically, trace theory purports to be offering a step-by-step causal account of memory. The ability to remember a piece of music (or anything else) is explained by the operation of a mechanism very much like a tape recorder. Perhaps the clearest account of this step-by-step causal process is offered in Wolfgang Köhler's principle of psychophysical isomorphism. On that account (as the reader will recall) every experienced feature of the music (e.g., loudness, pitch, rhythm, etc.) is causally linked to corresponding features of brain processes. These brain processes are in turn the result of interactions between the sound waves and the ear. When the music is remembered (called to mind), then it is assumed that the brain is producing impulses isomorphic to the ones caused by the sound waves when the music was actually heard. And trace theory, of course, explains this production of these patterned impulses by postulating the existence of an ismorphic trace. It is crucial to the theory that the trace contain a corresponding feature for every structural feature of the music (and, of course, for every feature of the brain process corresponding to that music.) By no means could trace theory allow the possibility that the causal chain be broken. Every step must be causally determined by the one before.It would not do, for example, if the trace theorist admitted that the brain could spontaneously produce impulses isomorphic to '4b'. If these impulses get produced, then they are the result either of stimulation from the auditory nerve (in the case of hearing the '4b') or activation of the trace (in the case of calling to mind the '4b'.) To say that the brain could spontaneously produce those impulses without an isomorphic trace is just like saying that a tape recorder could produce a piece of music from a blank tape. And such a claim runs counter to very basic notions of causality. Only a magic tape recorder could produce the '4b' without a tape of it.

The examination of trace theory resumed with the observation that many people are quite capable of calling to mind pieces of music which are slightly different than the music which was originally heard. While at first this observation seemed to pose no problems for the trace theorist, it was subsequently seen that the trace theorist was forced to admit the possibility that the brain could reproduce some features of the music *without* those features being represented in the trace. Then it was observed that this power (the ability to call to mind pieces of music changed from their originals) had no limit. That is, many people can call to mind renditions of a particular piece of music so changed from the original that no aspect of its acoustic structure remains unaltered. In a case like this, the trace theorist would be forced to maintain that the brain can spontaneously produce (impulses isomorphic to) a highly complex and organized piece of music for which there is no trace. And if this is possible in the case where the piece of music is (e.g.) 'The

Fourth Brandenburg Concerto' with all sorts of changes made, then it is also possible in the case where what is called to mind is just the regular 'Fourth Brandenburg', with no changes made.

The trace theorist might wish to raise an objection at this point. He might say that my argument was not directed against trace theory as a theory of *memory*. For all that I have shown is that trace theory does not account for the ability to call to mind (or imagine) *altered* pieces of music. And what, after all, does this have to do with the ability to *remember* a piece of music? Perhaps there are two completely different mechanisms involved, requiring two sorts of explanations. That is, there would be a memory mechanism, described by the trace theory, and an 'imagination mechanism', presumably described by theories dealing with the phenomenon of imagination.

Although this objection seems important at first, it actually comes to nothing. The point of my arguments was not to prove that there are no such things as memory traces. At this preliminary stage of the attack, all that has been shown is that, by the basic assumptions of trace theory (especially the concept of the step-by-step causal 'transcribing' process, and the principle of psycho-physical ismorphism) it must be admitted that *some* mechanism in the brain is capable of producing impulses isomorphic to particular pieces of music, with no isomorphic trace to guide this mechanism. And if this mechanism can produce impulses isomorphic to '4b' played too fast, with trumpets, too loud, in rhumba rhythm, etc., then why can't this same mechanism produce impulses isomorphic to the normal '4b'? In other words, it is unimportant whether or not there are two separate mechanisms, one for imagination and one for memory. For, by using the trace theorist's own assumptions, we have shown that the imagination mechanism is capable of taking over the functions of the memory mechanism. Thus, whether or not there are separate mechanisms for imagination and memory, it has been shown that there *need* not be a trace. And this conclusion should be quite surprising. For the trace theory seems at first to be inescapably true. But now we have seen that, by the basic assumptions which underlie trace theory, it must be admitted that the brain often does produce (impulses isomorphic to) music without an isomorphic trace. This conclusion leaves us no longer committed to the *necessity* of a trace theory.

Some readers will surely have objections at this point. They may still feel uncomfortable with the suggestion that we are now freed from the necessity of a trace theory. Because this point is so important, I will ask the reader's indulgence while we reexamine the objection from a slightly different angle. We might imagine an objector saying the following: "When Bach composed the '4b' his brain (we shall suppose) produced impulses isomorphic to the '4b', even though there was no memory trace of the '4b' recorded in his brain. Bach's *imagination* produced the '4b'. But when *you* call to mind the '4b', surely a completely different process takes place. Rather than your imagination 'creating anew' the '4b', your mind simply *recalls* the trace of the '4b', a trace laid down in the past by whatever part of your brain is concerned with *memory*. By ignoring the distinction between these two completely different processes, (creative imagination on the one hand, and memory on the other) your reasoning goes astray. You may have brought up an objection against some theory concerning the *imagination*, but how on earth have you shown that we are freed from the necessity of a trace theory of *memory*? How have you shown that there need not be a trace?"

This objection is based on a misunderstanding which I will explain. When I said that we were freed from the *necessity* of a trace theory, I did not mean that we had shown trace theory to be false, or that we had proved that there are no memory traces. For all that I have said up to this point, we might find that some form of trace theory is correct. All that I have attacked is the idea that some form of trace theory *must* be true. Remember that we started with the feeling that trace theory was inescapably true. We felt that the existence of memory phenomena made it necessary that traces exist. But what we have now discovered, startlingly enough, is that this original feeling of iron necessity was an illusion. For, using the trace theorist's own assumptions, we find that the brain must be credited with the ability to produce complex tunes *without* a trace. And, of course, I am not referring to a case of extraordinary creative genius (Bach's composing), but to an ordinary case of calling some altered tune to mind. Since a person's brain can create complex tunes without a trace in one case (calling to mind the altered '4b'), there is no reason to think that the brain can't do the same in the other case (calling to mind the unaltered '4b').

So far, then, we have only seen that the brain might, or might not contain memory traces. Only in this sense are we freed from the necessity of a memory trace theory. However, we shall eventually see that no form of trace theory is acceptable. Before pushing the examination any further, though, I want to mention two other objections which might possibly be bothering some of my readers. Suppose I have convinced my objector that, on the trace theorist's own assumptions, an imagination mechanism exists which is capable of producing 'memory reactions' in the absence of a trace. And so, in this sense, we are freed from the necessity of a trace theory. Nevertheless, he or she may still say that, whether or not an imagination mechanism exists which is capable of producing 'memory reactions' in the absence of a trace, there are still *other* reasons for insisting on the necessity of a memory trace. First, someone might ask, "If there is no trace, what happens to the distinction between imagination and memory? What, then, would be the difference between my remembering the '4b' and my inventing it anew?" So there must be a trace, in order to account for the difference between imagination and memory. Second, someone might say this: "How would we ever know that we are remembering the '4b' correctly, unless we *did* have a trace, a copy of the '4b' with which to compare the present memory?" So a trace is necessary for this reason as well. In view of these objections, it seems that we are not freed from the necessity of a memory trace.

These objections are not negligible, and they will be discussed.* But they are misplaced, and will not be discussed right now. Why are they misplaced? The best way to show this is to point out that we are examining trace theory *from within*. That is, we are supposed to be taking the part of the trace theorist. Thus, when we say that we are freed from the necessity of the trace, we mean that, starting with the trace theorist's assumptions about the workings of the brain, we, as trace theorists, find that the brain apparently contains an imagination mechanism capable of (e.g.) taking over the functions of the memory mechanism in case of injury. This imagination mechanism is apparently capable of producing fully orchestrated 'music impulses' even in the complete absence of any sort of trace. Now, being good scientists, we reason in the following way: "Since this newly-discovered imagination mechanism is completely capable of producing phenomena such as the whistling of tunes, why shouldn't we discard the idea of a separate memory mechanism? Regardless of questions about the difference between memory and imagination, we see that this postulated imagination mechanism is powerful enough to make the memory mechanism superfluous. Therefore, although there may be a memory mechanism, there *need* not be one. Questions about recognition mechanisms, and questions about the difference between memory and imagination will be dealt with by other scientists." The point is that the trace theorist is looking at memory as a mechanistic phenomenon. So for him, the questions about memory reduce to questions like; "What kind of mechanism must the brain contain, in order to enable us to whistle the '4b'?" The theorist's first inclination was to answer that the mechanism must be a *trace* mechanism. What we have seen is that, by his own assumptions, the trace theorist should now admit that there need not be any trace mechanism. On *his own view*, the brain is constructed in a manner which renders a separate memory mechanism superfluous.

* See pp. 142–146.

Let us return to the objections which gave rise to the preceding paragraph. "If there are no traces, how could we distinguish between memory and imagination?" "How would we ever know that we were remembering the '4b' correctly, unless we *did* have a trace, a copy of the '4b' with which to compare the present memory?" For us to reply to these objections now would be for us to defend the position of the scientist in the preceding paragraph (the position of a trace theorist who found that some different sort of mechanism could account for memory phenomena). But now is the time to remind the reader that we have, until this point, been playing Devil's advocate. Taking the trace theorist at his word, we found that memory traces are, surprisingly enough, superfluous. This realization leaves the trace theorist confronting an unhappy paradox. He starts with the assumption that traces are necessary for memory. In the course of explaining just how traces are supposed to function, he ends up asserting that traces are not necessary for memory.

If the position to which the trace theorist has been led is open to objection, we may (somewhat heartlessly) choose just this moment to abandon his encumbered enterprise. Indeed, having led him this far, we will prove more treacherous still. For now we will turn around and attack the trace theorist's position ourselves. The attack begins like this. Our trace theorist may now say that, even if we don't need memory traces, still there may be such things. That is, even if there *need* not be separate memory and imagination mechanisms, nevertheless there still *might* be such mechanisms. And thus, though trace theory is not *necessarily* true, we may yet find that some form of trace theory is, in fact, true. Our reason for assuming that some form of trace theory is correct is a strong one. The most important feature which makes the theory attractive is the fact that it offers an explanation of memory which is hard-nosed and scientific. Each step in the memory process is part of a straightforward causal sequence. And this sort of account supposedly gets its strength by comparison with other sorts of accounts of memory. A theory of memory which demanded some sort of spiritual soul as a requirement would rightly be dismissed as unscientific, for example. Such a theory would be considered mystical by sympathizers and would be considered nonsense by others. But trace theory claims to avoid any such charge (i.e., the charge that it is mystical nonsense or supernatural mumbo-jumbo.) But is this correct?

What has been shown in this chapter is that there need not be a memory trace. Indeed, even if there are traces, the theorist must admit that many people can call to mind a piece of music for which there is no trace (e.g., the example of '4b', played too fast, with trumpet, etc.). And so, the trace theorist must admit that, on his own view (especially the principle of psychophysical isomorphism) the brain is capable of producing highly organized impulses (as in the example just mentioned) without any trace. And once the trace theorist admits that this is possible, then he has violated the very heart of his theory. That is, he must admit that the brain can produce these highly organized impulses *spontaneously*. And this breaks the chain; this destroys the step-by-step causal 'transcribing' process. But the conviction that there *must be* some sort of step-by-step causal process is the chief motive for postulating a trace theory of memory. This sort of theory is supposed to be a way of demythologizing the powers of the mind. And yet it seems as if the trace theorist is left with just the sort of assumptions he wished to avoid. In particular, the trace theorist is forced to say that there is a machine (the super-computer known as the human brain, which includes a recording device) which can produce music in the absence of any sort of recording of that music. But the only sort of machine which could do that is a magical, mystical, supernatural machine.

Once the trace theorist allows the causal chain to be broken at one point, the whole chain naturally collapses. If the machine is given mystical powers in order to explain how it performs some particular function, then the explanation loses its scientific character. One might as well say that the machine has mystical powers when it performs other functions as well; this would not make the explanation any more mystical. If you believe that one ghost exists, then you believe in ghosts. Suppose the following question is asked: "Why is the trace theorist (a *memory* theorist) responsible for the shortcomings of a theory of *imagination*?" The answer is: It is the trace theorist's own mechanistic assumptions which leave him *committed* to the existence of an imagination mechanism of the type described (that is, one which works by magic).

At this point it would be a good idea to clear up one possible source of confusion. Perhaps the example of a person who 'hears a piece of music in his mind', or 'in his head' is causing trouble, or provoking suspicion. So I would like to point out that there is an easy way to avoid any confusion raised by the example of someone hearing something 'in his head'. Instead, we can substitute the example of someone humming out loud, for all the world to hear. This type of example would do just as well; the point of the examples is to show that people can *produce* the various versions of the '4b' (or some other piece of music) at will. And by the assumption of the principle of psychophysical isomorphism, the brain must be producing impulses isomorphic to these various renditions of '4b'. The shift from 'hearing in the head' to humming is permissible because there are people who are very good at imitating the sound of various instruments. This power or ability is perhaps not so widespread as the ability to call to mind a piece of music. But it is certainly not unusual. And when we encounter someone with this ability we may be interested or amused; but we certainly do not regard it as magical. So let's just assume that we are dealing with someone who can imitate various instruments. For example, I myself can do a fair imitation of a trumpet or a string bass playing the opening bars of the '4b'. It is not essential to my examples that they take place 'in the mind'.

The problems which have been raised so far are not peculiar to cases of *hearing*. The same difficulties arise when we examine a typical trace theoretical explanation of how we remember things that we have *seen*. Let us consider the case of someone who has been to a museum and seen a painting (the 'Mona Lisa', for example.) There is a background of accepted scientific theory which would be assumed by the trace theorist (and anyone who had been educated in our society.) The assumptions being referred to are those of classical optics. They would include explanations of such terms as 'wavelength', 'focus', 'lens', 'color', 'retina', 'optic nerve', 'refraction', and so on. These theoretical assumptions correspond, of course, to the assumptions of classical acoustics, which formed the necessary background for the example of remembering the '4b'. To avoid undue length, we'll pass by the full-blown explanation of the optical phenomena which occur when someone looks at a painting. Let's just say that the light strikes our eyes and causes nerve impulses to travel down the optic nerve to the brain. In a similar fashion we can simplify the trace theorist's account by supposing that the brain contains something like a camera or a videorecorder. And so trace theory would offer an explanation of the following sort: When someone remembers a painting (when he 'sees it in the mind's eye'), it must be that a trace of the painting is being activated somewhere in the brain. So, for example, when someone remembers the Mona Lisa, when someone 'pictures the Mona Lisa' to himself, the face of the lady in the painting is seen as flesh colored. Not everyone has this ability to call up vivid visual images; but many people do. And the trace theorist assumes that the flesh color of the face must be somehow represented in the trace of the Mona Lisa. Otherwise people might remember her face as being any random color – blue, for example. And this would be true of every optical feature of the image. For example, the relative size of the Mona Lisa's head must be represented in the trace. Otherwise, when people remembered the Mona Lisa they might remember her head as being wider than her shoulders, or perhaps as being no bigger than an orange. We can sum this up by saying that the trace of the Mona Lisa must be completely isomorphic to the painting, down to the smallest detail. (That is, down to the smallest detail that the particular person can correctly recall.)

Once again, as in the case of remembering the '4b', an isomorphic trace seems to be absolutely necessary. How else, for example, could we explain the fact that the face is flesh colored in the memory image? In this example, the nervous system (the brain) is acting as a projector which somehow manufactures an accurate image of the Mona Lisa. Now we know of machines which can do this. A videorecorder transforms magnetic patterns (which have been previously recorded onto a tape) into patterns of light. These patterns of light are the images we often see on a television screen. But can anyone imagine a videorecorder which manufactured images without the help of some sort of previous recording? Of course we can imagine machines which manufacture patterns of light, even without previous recordings. But such machines do nothing but generate random patterns. The machine *we* are interested in, however, is one which can generate a picture of the *Mona Lisa*. And, of course, the only sort of machine which could do that without a previous recording is a magical machine – a sort of ghostly slide projector which somehow creates a picture of the Mona Lisa out of thin air – and then manages to get the color right! Since the trace theory purports to offer a *scientific* explanation of the power of creating visual images, magical explanations are not to be tolerated.

The necessity of an isomorphic trace seems ironclad. The only alternative seems to be to give up the search for a scientific explanation and embrace a mystical one. Whether there really is an exhaustive dichotomy here is a problem for future discussion.*

Our examination of this example begins the same way as our examination of the musical example. Think, for instance, of how the Mona Lisa would look with a blue face. Many people are quite capable of visualizing this with no trouble. As in the acoustic

* See pp. 146–150.

example, the trace theorist at first seems to have little trouble explaining this phenomenon in terms of trace theory. All he need say is that the visual trace of the Mona Lisa is being activated in a way slightly different from the usual one. Perhaps the trace is being activated and a color generating mechanism simply changes the flesh tone to a blue tone. After all, we can do this with a color television; and the brain is a million times more complex than any television could ever be. So where, one might ask, is the problem? But this move gets the trace theorist in trouble, just as it does in the acoustic example. Consider: the trace theorist is saying that the brain is quite capable of producing an image of the Mona Lisa with a blue face, even though the original memory trace bears no indication of any such color in the face. That is, whatever feature of the trace corresponds to color, (or face color) surely this feature would be the one for flesh color, not blue. Yet somehow the brain creates an image of a blue-faced Mona Lisa. So even though there is no indication of blue color in the trace, nevertheless the (hypothetical) color generating mechanism steps in and puts in a blue color where the trace has an indication for flesh color. If this is possible, then for all we know, and for all the trace theorist knows, the very same sort of process could occur in the case where someone visualizes the Mona Lisa with a flesh colored face. The trace of the Mona Lisa might, e.g., contain a green color indicator for the face. And when I (or whoever) visualize the Mona Lisa with a flesh colored face, the color generating mechanism steps in and puts in the flesh color rather than the green color indicated by the trace. And, of course, this possibility applies not only to the face of the Mona Lisa, but to any part of the image. And further, there's nothing special about the image of the Mona Lisa. The trace of *every* visual image might contain a green color indicator. And whenever I visualize something, the color generator simply puts in the proper colors. Maybe the color indicators of memory traces are in black-and-white. Or maybe just black. And that is to say that there might be no indication of color at all; the color generator just puts in the right colors, with no color indicators at all.

The reader quite possibly can guess the next step in this chain of reasoning. Various other optical features of the image can be varied at will. Not only can different colors be imagined for the Mona Lisa's face, but (e.g.) different sizes for the head. I can easily produce an image of the Mona Lisa with a very small head. In this case, the trace theorist assumes that (say) a size-reducing mechanism acts on the normal head-size-indicator of the trace. But, of course, the same sort of process could go on when I visualize the Mona Lisa with a normal size head. That is, the size-reducing mechanism could act on an oversize-head-indicator, thus producing an image with a normal size head. In other words, the trace of the Mona Lisa might very well contain a head-size-indicator larger than we would have expected. And yet, when I visualize a normal Mona Lisa, the size-reducing mechanism does its job. And, of course, the trace might instead contain a head-size-indicator smaller than we expected. (It might, for example, show the Mona Lisa with a head the size of an orange.)

If the size-reducing or size-enlarging mechanism can do this with the Mona Lisa's head, then it can do the same with any part of the visual image. Imagine standing in awe before a mile-high portrait of the Mona Lisa. Or imagine peering with admiration at a tiny miniature of the Mona Lisa. And there is nothing special about the image of the Mona Lisa. Any optical trace could exhibit these peculiarities. Maybe *all* optical traces bear size-indicators for images of the *same* size. And when these traces are activated, the enlarging and reducing mechanisms do their jobs. So the trace of a battleship image is enlarged and the trace of a gnat image is reduced when they are activated (when someone visualizes a gnat or a battleship.) But this is tantamount to admitting that the trace need bear no indication of size at all. Put it this way: if every road sign on the New York Thruway said 'NEW YORK – 10 MILES', then the signs wouldn't tell you anything at all about how far it is to New York.

The situation is completely analogous to the acoustic example. Any feature which can be changed in imagination is shown, by the same process of reasoning, to be inessential as a part of the trace. We can (at least I can) easily visualize the Mona Lisa with a moustache, with an orange-sized head, or even with an orange for a head. Whatever mechanism generates these monstrosities is quite capable of generating the usual image of the Mona Lisa. So why do trace theorists assume that there must be a trace in the latter case? The trace theorist is forced to assume the existence of a brain mechanism capable of producing impulses isomorphic to a particular painting (e.g., the Mona Lisa with a huge orange instead of a head) even in the absence of a trace of such a painting. He is committed to the existence of this mechanism in the following way: (1) I *can* visualize this monstrous Mona Lisa. (2) I have no memory trace of it, since I have never seen such a painting. And (3) by the principle of psychophysical isomorphism, every aspect of the visual image is (must be) matched by a corresponding feature of some brain process. As we remarked before, the only sort of mechanism which can create complex pictures (e.g., the Mona Lisa with a huge orange for a head) out of thin air – is a magic mechanism.

NOTES

[1] Wolfgang Köhler, *Gestalt Psychology, op. cit.*, p. 37.
[2] *The Task of Gestalt Psychology*, by Wolfgang Köhler, Princeton University Press, 1969, 164 pp., p. 66.
[3] *Gestalt Psychology, op. cit.*, p. 39.
[4] *Gestalt Psychology, op. cit.*, p. 140.
[5] *Gestalt Psychology, op. cit.*, p. 141.

PART II

BROADENING THE ATTACK

INTRODUCTION

In the preceding part of this work, the reader was introduced to trace theory. An example was then put forward – one which created problems for the trace theorist. In the second part of the book, our examination will be broadened to include two very popular variants of trace theory. These two are the 'stimulus-response' and the 'information processing computer' trace theories. They come to be included in our examination in the following way. In the first chapter, trace theory will be shown to be subject to another criticism – one which seems to destroy any claims to plausibility which the theory might have enjoyed. In the second chapter, we will examine the two variants. At first glance, it might be thought that they enable trace theory to avoid the criticism. But we shall see that this success is only apparent; both avoidance maneuvers fail, thus ending the second part of the book.

CHAPTER ONE

ANOTHER PROBLEM FOR TRACE THEORY

For trace theory to be at all plausible, it must not only explain how memory traces are *recorded*, but it must also explain how these memory recordings are *played back*. The problem I have in mind is best illustrated by an example. Suppose that Mr. A has learned the 'Star Spangled Banner'. He was brought up in the United States, has attended public school here, etc., and so he can now hum the national anthem. The trace theorist supposes that Mr. A has a recording of the 'Star Spangled Banner' somewhere in his central nervous system – probably in the brain. When Mr. A hums the tune on some particular occasion, the start of a baseball game, for example, the trace theorist assumes that Mr. A's recording of the 'Star Spangled Banner' is somehow activated, or 'played back'. This assumption is, of course, vital to trace theory. After all, the reason that the traces are laid down in the first place is to provide some sort of model of things past. In this example, the musical trace is resorted to in order to explain how Mr. A gets the tune right. He doesn't just get it right by accident. And in order for the trace to be of use, it must not only be recorded, it must also be played back.

> ... in the sequence of events constituting an act of memory ...... there are three such events: input of information into the memory store, storage of information (i.e., maintenance of information in the store) and retrieval of information from the store.[1]

The problem of playback, or retrieval mechanisms, is one toward which a great deal of attention has been paid. Current theorists are wrestling with such questions as: Is there one retrieval mechanism for all memory? Or are there perhaps different retrieval mechanisms for different sense modalities, or for short and long-term memories? The passage just cited, in fact, was taken from an article in which the author opts for a memory mechanism with two separate retrieval systems – one for short-term and one for long-term memory. Retrieval is an important problem for this reason: the retrieval mechanism must choose between an extremely large number of memory traces. Consider the various sounds that most people can identify. It takes no unusual talent to be able to identify the sounds of (e.g.) typewriters, pianos, seagulls, corks popping, trucks, ocean waves, thunder, heartbeats, churchbells, ice tinkling in glasses, doors closing, etc. And people with any musical ability can hum literally hundreds of different tunes. It would probably be better to pass over the innumerable commercial jingles that the average American has been exposed to. Unfortunately, people with musical ability are all-too-familiar with these commercial monstrosities, and can unhesitatingly hum or sing them, or recognize them when someone else does so. Add to this the fact that most of us can recognize quite a few voices of different people. Some of us can also imitate some of these voices. The point of producing this *extremely* fragmentary, casual list is to indicate to the reader just how many traces would have to be stored away in the brain. For the trace theorist assumes that the ability to recognize or reproduce tunes (sounds, sights, or anything else) is to be explained by trace retrieval.

The term 'memory' is often used in a very general sense. We will distinguish between *learning*, (his italics) a process which establishes traces in the nervous system, and these traces themselves, which form the basis for memory in a narrower sense. And both have to be distinguished from such facts as actual *remembering*, (his italics) which may mean either recalling or recognizing, both obviously being effects of traces on present psychological processes.[2]

What does recognition mean? It means that a present fact, usually a perceptual one, makes contact with a corresponding fact in memory, a trace, a contact which gives the present perception the character of being known or familiar. But memory contains a tremendous number of traces, all of them representations of earlier experiences which must have been established by the processes accompanying such earlier experiences. Now, why does the present perceptual experience make contact with the trace of the *right* earlier experience? (his italics) This is an astonishing achievement.[3]

Köhler is correct in saying that this is an astonishing achievement. In fact, the hypothetical mechanisms which trace theorists invoke to explain this achievement are unequal to the task. All of the variants of trace theory which I have examined, some put forth by researchers and theorists who show considerable sophistication, fail miserably in the vital matter of trace retrieval. I don't find this surprising, since I believe that such a failure is unavoidable. But what I do find surprising is that, among trace theorists, there seems to be a general blindness to this failure. The problem seems to come up at the same point in all the theories. What this problem is will become clear when we look at a few examples. In an article referred to a few pages back, we find this account of just how the retrieval mechanism is supposed to work:

Recall performance does not only depend on the information *available* (his italics) in the store, but also on *accessibility* (his italics) of that information. Accessibility of stored information is determined by retrieval cues available to the subject at the time of attempted recall . . .
. . . Retrieval from this unitary store occurs as a consequence of availability of retrieval cues which bridge the gap between the present environmental demands and the information stored in memory on an earlier occasion. Unless at least one retrieval cue exists for a given unit of information, the unit cannot be retrieved.
What events constitute retrieval cues that provide access to the information about TBR (this is the author's abbreviation for 'to-be-remembered') items in the memory store? In general, the nature of effective retrieval cues is determined by the coding of input material at the time of input. When a TBR unit is stored, some ancillary information about it is also stored with it. The storage of this ancillary information represents what is referred to as 'coding'. When some of this ancillary information (or the 'code' of the TBR unit) is available at the time of the attempted recall, the code serves as a retrieval cue. The effectiveness of retrieval cues thus depends, among other things, upon coding operations that have taken place at input and the availability of information about these coding operations at output.[4]

Now we see how, in one example, the trace retrieval mechanism is supposed to work. If we think of the trace as, say, a tape recording, then the retrieval mechanism is faced with a large number of tapes to choose from. How does it pick the right one? Tulving says that, for each 'TBRU' (and this chapter is not the place to ask what sort of 'units' he might have in mind) there is at least one retrieval cue. Thus, the mechanism is not working at random. That is, when someone is asked to hum the 'Star Spangled Banner', it is no mere accident that the person asked does so. And, of course, any memory theory which *did* make it a matter of accident, or random chance, when someone remembered something correctly, would be a ridiculous parody of a memory theory. Imagine: "Our theory states that each of the thousands of things that a person can remember is represented by a tape recording. And, when someone is asked to recall a particular item, it is purely a matter of random chance which tape recording is played back." That won't do. Tulving's theory is, on the other hand, typical.

But does Tulving's retrieval mechanism do the job? He says that there are 'retrieval cues' for each trace. The retrieval cues, says Tulving, are actually in the form of 'ancillary information' about the items they represent. Tulving perhaps has something in mind like the following: If the trace storage area is likened to a library, then the retrieval cues are cards in the catalogue. If the retrieval mechanism is instructed to recall the 'Star Spangled Banner', for example, then it does so by means of the card catalogue. Presumably, there would be a little card which stated that the 'Star Spangled Banner' was a national anthem, and was to be found on shelf # 63020, or something of the sort. I think this is a reasonable way to interpret Tulving on this point. But the retrieval mechanism described here could never work. Suppose that the retrieval mechanism is sent into the library to do its job. That is, it must pick out the tape of the 'Star Spangled Banner'. This tape, of course, is only one among a vast array of memory traces. So the retrieval mechanism must separate the right tape from all the others. Tulving's suggestion is that there is another 'ancillary' trace which is somehow 'coded' to match the tape of the 'Star Spangled

Banner'. We may think of this code as a sort of *key*. Each item of the code stands for some feature of the thing to be remembered. That is, the code contains features which correspond, in our example, to 'national anthem', 'United States', etc. Now, for each of these features, the 'key' in the card catalogue bears some sort of stamp. Pretend it's an actual brass key. Tulving's idea is that the tape of the 'Star Spangled Banner' has a keyhole. The lock inside the keyhole is the only lock which fits the key. That is, each memory trace bears its own, unique lock. And the card catalogue is a series of keys, one for each lock. Then we see that the retrieval mechanism merely runs down the shelves, holding the key in front of it. Each lock is tried in turn. When it comes to the trace of the 'Star Spangled Banner', the key fits the lock. And that's how, on Tulving's view, the retrieval mechanism works. Of course, we were using an extremely crude mechanical analogy. His own theory would involve very sophisticated codes and very complex mechanisms – not brass keys with bumps on them. And, of course, Tulving is speaking of memory traces, not reels of tape. Further, machines can be built which can do this sort of thing *very* rapidly (computers, for example, can be programmed to search through extremely long lists in an incredibly short time.) So it is no problem for the retrieval mechanism to run through the whole library in a fraction of a second.

But now, will this mechanism do the job? One very important step has been overlooked. The retrieval process described took place *after* the right 'card' had been taken from the catalogue, *after* the right key had been chosen, *after* the right 'ancillary' trace had been chosen. (These three are equivalent.) How did the retrieval mechanism manage to pick the right key, the right ancillary trace? How did it pick out the one with the right code, the only one with the code that matched the trace of the 'Star Spangled Banner'? After all, the catalogue must contain at least as many cards as there are traces. The key ring must have at least one key for each lock. (If there were fewer keys, then one key might open the locks on several tapes.) So how does the retrieval mechanism happen to pick just the right key, the right card, the right ancillary trace, from all of the thousands of others? This, of course, cannot be a matter of random chance. The retrieval mechanism could no more pick the right key by chance than it could pick the right trace by chance. In Tulving's terms, if the retrieval mechanism needs some 'ancillary' trace to pick out the right trace, then it must also need a second-level 'ancillary' trace to pick out the first ancillary trace. In other words, Tulving is faced with an infinite regress of card catalogues, keys and locks, or ancillary traces. Further, Tulving's retrieval mechanism is typical of those put forward by trace theorists. We shall examine some others. But before we do that, we must consider the possibility of alternatives to the infinite regress of traces.

What if Tulving objected that his retrieval mechanism did *not* need a second-level ancillary trace in order to pick out the first-level ancillary trace? Would this get him out of trouble? Well, it would save his theory from an infinite regress. But consider: If the retrieval mechanism is able to pick out the right *ancillary* trace without some further trace to guide it, then why can't the same sort of mechanism pick out the *actual* trace in the first place, without resorting to a guide, a code, an ancillary trace? The notion of a code, a guide, or a key does absolutely no work here. At some level the machine *must* pick the right trace, without a guide. So Tulving's account of retrieval fails, and with it, his trace theory. Further, the considerations which lead us to reject Tulving's solution are general. That is: all trace theories require a retrieval mechanism. The notion of second-level traces or coded traces is completely empty of explanatory value. So any trace theory which postulates that kind of retrieval mechanism (i.e., one which is guided by ancillary traces or codes) is a failure.

We are now in a position to see that, in order for trace theory to provide an explanation of human memory, there *must* be a retrieval mechanism which can pick out the right trace without any sort of guide. Accordingly, the question as to whether there can be such a mechanism is of prime importance. Let us assume, for the sake of argument, that we do have a mechanism which is capable of picking out the right trace, and that this mechanism is *not* guided by some key or code which matches the right trace. What would be required of such a machine? It would have to be able to pick out the right trace from a great number of wrong ones. But how would the machine know where to stop? How would it recognize which was the right trace? Suppose, for example, that the machine set out to fetch the trace of the 'Star Spangled Banner'. How would it know which was the right trace? Remember, please, that the machine can't do it by matching the trace with some key. We ruled that out as a possible explanation. So how on earth can the machine fetch the right trace? What sort of process could possibly make it choose the trace of the 'Star Spangled Banner'? Why shouldn't it equally well stop and choose the trace of 'America the Beautiful'? Or, for that matter, the trace which represents the sound of an atomic explosion? What kind of mechanistic process (aside from matching) could possibly account for this extraordinary performance on the part of the machine?

A machine which *just stops* in front of the right tape, or a machine which *just chooses* the right trace, is a machine which knows where to find it, or knows what to look for. If there is nothing to guide the mechanism, and yet it picks out the trace of the 'Star Spangled Banner', then that 'mechanism' just knows what it is looking for; it must know the 'Star Spangled Banner'. And that means it must have learned what the 'Star Spangled Banner' is, or what it sounds like. And that means it must be a machine which knows, thinks and remembers. And, finally, what that means is that the trace theorist has failed to analyze memory (or anything else) into mechanistic terms. All that has been done is to attribute to the retrieval mechanism the very power that was denied to people at the outset: Memory. The idea of a machine which knows, thinks, or remembers, is the idea of the little black box with the homunculus inside. And a theory which requires the homunculus is no explanatory theory at all. At least, it is no scientific theory. Imagine:

A) "How do people remember things?"
B) "They have machines which do the remembering."
A) "But how do the machines work?"
B) "Simple; the machines have little people inside."

Since the trace theory requires a retrieval device, and since the retrieval device requires a homunculus to operate it, I conclude that trace theory is a fruitless attempt at a scientific, or causal, explanation of memory.

The trace theorist might object that he does not have in mind a black box with a little man inside. Rather, he means that the traceless retriever is a bona fide *mechanism*. But we have encountered this sort of mechanism earlier. The mechanism that just knows which is the right trace, the mechanism that just remembers (with no trace to guide it) which is the trace of the 'Star Spangled Banner', is the same sort of mechanism that can spontaneously, without guidance, produce a full-blown rendition of the 'Star Spangled Banner', or that can spontaneously create an image of the Mona Lisa. Suppose that somehow, some machine *did* spontaneously create an image of the Mona Lisa, but with a blank space above the shoulders, no head at all. Now suppose that the machine is about to put on a head – it is about to complete the picture. Remember, the machine does not have anything to tell it what to do. And somehow, it just spontaneously puts the Mona Lisa head into the picture. A machine like this is one which works according to magic, not science. How would it come to put on the right nose, for example? It would be a miracle if the machine put in a human nose rather than a radish, for example. And how would it come to color the human nose with the proper shade of flesh? Why not dead white? Or bright red? Or red, white, and blue? This mystical, supernatural machine just happens to put in the right nose, with nothing to guide its action. My point here is that a machine which puts on the right nose without guidance is no more incredible or nonsensical than a machine which just picks out the right trace, with nothing to guide it.

This is, I believe, a very important point. It will not do to have a retrieval mechanism that is guided to the right trace by means of a code or key. That results in an infinite regress. So the trace theory *demands* a traceless retrieval mechanism. Yet such a mechanism is every bit as nonsensical as a mechanism which, somehow, with nothing to guide it, puts a Mona Lisa head on a Mona Lisa torso.

If the trace theorist admits to the necessity for a homunculus, his theory is a sham explanation. For the question, How does the little man in the black box remember anything? is the same question as, How do people remember anything? The homunculus does not provide us with an explanation; a fortiori, it does not provide us with a scientific, causal or mechanistic explanation. Alternatively, if the trace theorist opts for the unguided retrieval mechanism, then he has attributed magical powers to his machine, And that doesn't sound like a scientific, causal or mechanistic theory, either. For if the machine requires magic to perform even one function, then it is a magical machine. To repeat a remark made earlier, if you believe that one ghost exists, then you believe in ghosts.

NOTES

[1] Page 7 of the book, *Biology of Memory*, edited by Karl H. Pribram and Donald E. Broadbent, Academic Press, 1970, New York, London, 323 pp. From the article entitled 'Short and long-term memory: different retrieval mechanisms', by Endel Tulving, University of Toronto.

[2] Wolfgang Köhler, *The Task of Gestalt Psychology, op. cit.*, p. 121.

[3] Wolfgang Köhler, *The Task of Gestalt Psychology, op. cit.*, p. 122.

[4] From the article by Endel Tulving, p. 8 of the book *Biology of Memory, op. cit.*

CHAPTER TWO

STIMULUS-RESPONSE AND INFORMATION PROCESSING COMPUTER THEORIES OF MEMORY

Our examination of trace theory has essentially been concluded. At this point it should be clear to the reader that no trace theory can provide a satisfactory causal account, or mechanistic explanation, of memory. It should be remembered, however, that trace theory was seen to be the only possible candidate for a mechanistic explanation of memory. The alternative to a trace theory is a traceless theory. And, since we are talking of mechanistic theories, this traceless theory is the theory of a *traceless memory mechanism*. But, of course, we have seen that this is a nonsensical concept. Once again, we have here the idea of a machine which can (e.g.) paint a picture (an image) of the Mona Lisa, with nothing to guide each step. And as each feature is added to the picture, we can ask, Why did the mechanism just put two eyes on her head? Why not three, or one, or none? If it just does these things spontaneously, with no guiding pattern, then it is a magic mechanism. The argument is the same, whether we are dealing with a traceless retrieval mechanism, or a completely traceless mechanistic account of memory. The machines in both cases are required to do the same (impossible) thing. So, when we attacked trace theory, we were also engaged in an attack on mechanistic theories of memory in general.

Our conclusion can, perhaps, best be put in the following way: Any theory which purports to be a causal, or mechanistic, or scientific account of memory, will require the existence of a homunculus, a magical machine, or an infinite regress of mechanisms. And any such theory would be neither causal, nor mechanistic, nor scientific. In this chapter we will examine stimulus-response and information processing theories of memory. Both sorts of views are to be considered as variants of trace theory. While the reader will be introduced to the two theories, no attempt will be made to characterize them in the same detailed manner as we characterized trace theory in general. Rather, we shall probe just deep enough to expose the homunculus hidden in both theories. In other words, we shall examine them long enough to show that they are not exceptions to the conclusion stated at the beginning of this paragraph. The only reason that we will examine them at all is their great popularity among today's scientists and scientifically-minded laymen.

We shall examine stimulus-response theories of memory first. A good way to start would be to see how B. F. Skinner defines the two terms 'stimulus' and 'response'.

> We need to establish laws by virtue of which we may predict behavior, and we may do this only by finding variables of which behavior is a function. One kind of variable entering into the description of behavior is to be found among the external forces acting upon the organism. . . . The environment enters into a description of behavior when it can be shown that a given *part* (Skinner's italics) of behavior may be induced at will (or according to certain laws) by a modification in part of the forces affecting the organism. Such a part, or modification of a part, of the environment is traditionally called a *stimulus* (Skinner's italics) and the correlated part of the behavior a *response*. (Skinner's italics)[1]

Skinner sees himself as attempting to provide an analysis of human behavior into mechanistic terms. It is his aim to eliminate, by means of his analysis, any notion of a free agent or will. While he himself does not claim to carry out this analysis to the level of molecules or neurons, he does not in principle rule it out.

> The work (i.e., Skinner's book) is 'mechanistic' in the sense of implying a fundamental lawfulness or order in the behavior of organisms, and it is frankly analytical. It is not necessarily mechanistic in the sense of reducing the phenomena of behavior ultimately to the movement of particles, since no such reduction is made or considered essential; but it is assumed that behavior is predictable from a knowledge of relevant variables and is free from the intervention of any capricious agent.[2]

We will not examine Skinner's views *per se*. Rather, we will be concerned with those views which use Skinner as a starting point. While Skinner himself is not concerned (at least here) with reducing human behavior to the behavior of neurons, his concept of stimulus-response has been used by a number of memory theorists. It is precisely their concern to bring about the reduction of human (memory) behavior to this level of analysis, thus eliminating the need for such unscientific concepts as the thinking, perceiving, willing agent. These theorists make use of Skinner's concept of stimulus-response in something like the following way. Suppose someone (Mr. A) has learned the 'Star Spangled Banner'. Let us simplify this by supposing that somehow, a trace of the 'Star Spangled Banner' has been laid down in Mr. A's central nervous system. When, for example, he is attending a baseball game, and the crowd begins to sing the 'Star Spangled Banner', the stimulus of the crowd's singing evokes the response in Mr. A. The stimulus causes recall of his trace of the 'Star Spangled Banner', and Mr. A sings along. This kind of account fits Skinner's definition. The sound waves put forth by the singing crowd would constitute the 'modification of the part of the environment' – the stimulus. And the recall of the trace, including Mr. A's subsequent singing, would be the 'correlated part of the behavior' – the response. In his book, *Brain Memory Learning*, for example, W. R. Russell defines memory as "the capacity to repeat previous responses".[3] Since the response in this case is the singing of a piece of music, it is clear that a stimulus-response theory of memory will require a trace. Otherwise, of course, it would be impossible to explain why the organism responded with (e.g.) the 'Star Spangled Banner'.

Stimulus-response theorists are not, for the most part, interested exclusively in theories of memory. Many are concerned to offer more-or-less complete accounts of human behavior in terms of Skinner's principles. Some, for example, are interested in memory as a manifestation of *learning*. They see themselves as offering *learning theories*, rather than memory theories. For them, learning is the capacity to change the response to a given stimulus. An example of such a change is embodied in the old saw: the burnt child fears the fire. An analysis of this example in terms of stimulus-response learning theory would go something like this. A child sees a fire (stimulus). The child reaches for the fire (response) and gets burned. The next time the child sees the fire (a repeat of the stimulus) it does not reach for it. The child's response to this particular stimulus has changed. This is an example of learning. Should the child meet with the fire again, it avoids it, having learned that the fire burns. When the child displays this changed response to the repeated stimulus, it is displaying the faculty of memory. Learning, then, is the process whereby different stimuli are paired with various responses. So the stimulus-response learning theorist is investigating the process by means of which particular 'memory responses' are set up. It is this close link between the two which accounts, I think, for the fact that a number of stimulus-response theorists seems to use 'memory' and 'learning' in the same breath.[4]

This type of stimulus-response view seems to offer a great advantage over other types of trace theory. The advantage lies in the area of trace retrieval. Here at last we are provided with a feasible retrieval mechanism. Retrieval is accomplished by purely mechanistic, automatic processes, requiring no magic to make it work. The question, "How does the right trace get picked out?", has an answer. It is the stimulus itself which accomplishes retrieval of the correct trace. There is a direct mechanical (or 'biomechanical') connection between stimulus and response. We may compare this sort of retrieval with the retrieval of a record disc by a juke-box. The right combination of buttons pressed causes the mechanism to move down a predetermined pathway, stopping at the right record. In the same way, the right combination of stimuli (e.g., sights and sounds) causes the playback of some trace.

At this point, we would be forced to admit that this sort of retrieval mechanism could do the job. Or at least there is no philosophical criticism of the theory behind this kind of retrieval mechanism. Any objections to the stimulus-response retrieval mechanism would have to come from experts in the fields of biology, or biochemistry, or biophysics, or some other related field. Since we (I) have no training in these fields, it would not be appropriate to try and raise and objections. Once the connection is forged between some particular stimulus and some particular response, retrieval of the trace (the response) becomes automatic. This is the sense in which stimulus-response theories represent an advance over the trace theories we have examined so far.

But now it must be noted that we have not yet described a complete memory mechanism. True, retrieval becomes automatic, *once the connection between a particular stimulus and a particular response gets established*. But this setting up of connections between stimulus and response must be a very complex process. In fact, all of the problems which cropped up when we considered retrieval mechanisms return to plague the stimulus-response theorist under the guise of problems of *learning*. For the setting up of connections between stimulus and response is precisely what stimulus-response theorists call 'learning'. In our example of the burnt child, learning was constituted by the change of connections between stimuli and responses. Specifically, a new connection was set up between the stimulus of the fire and the response of avoidance. And an old connection was broken (or at least altered) between the stimulus of the fire and the response of reaching out. Now, if the stimulus-response theorist can come up with a plausible account of how these connections come to be set up, his theory is in a strong position. However, if the theorist cannot plausibly explain how particular stimuli come to be connected with particular responses, then his account fails. And the failure of this part of his theory will result in the failure of the stimulus-response account of *retrieval*. For, as we have just seen, the retrieval mechanism postulated earlier was one which could work only on condition that connections exist between particular stimuli and particular responses.

("How come the retrieval mechanism stopped at just that trace?" Answer: "A connection exists between that stimulus and that trace, that response.") Accordingly, let us examine this process of connection between stimulus and response.

This process of linking stimuli with responses is generally called 'association'. James G. Greeno offers a clear statement of the central part played by this association process in any reasonably plausible stimulus-response theory.

Learning theory as we know it today probably was founded in the 17th-century, when Hobbes and Locke revived Aristotle's attack on the doctrine of innate ideas. Hobbes and Locke and other empiricist philosophers took the view that knowledge comes from experience. This view requires a learning mechanism, and the empiricists proposed that learning is a process of combining impressions that occur near one another in space and time, or are similar, or contrast with one another. . . . It seems safe to say that the belief in association as the elementary learning event has dominated theories of learning and thinking for at least three centuries. The early view that associations form between ideas has been replaced in this century by the idea that associations connect stimuli and responses.[5]

I propose that the first stage of memorizing and association involves storing a representation of the stimulus-response pair in memory as a unit.[6]

The central importance of association is clearly shown by the attention paid it by memory theorists. Wolfgang Köhler, for example, treats extensively of the association between a present experience and a trace.[7]

The proposed process of association, however attractive it may seem, will not explain the process of trace retrieval. A mechanism of association could only perform under an extremely simple set of conditions. And, unfortunately for the stimulus-response theorist, the phenomena he sets out to explain are anything but simple. More precisely, the problem is this. If the retrieval of appropriate responses (or the traces which bring about these responses) is to be satisfactorily explained, there must be an extremely limited number of connections between a given stimulus and its associated responses. If the association can be represented by a one-to-one mapping of stimuli to responses, then there is no problem. But if a given stimulus can call forth more than one response, the theorist is faced with a challenging problem. Suppose the stimulus is a photograph of a friend. If humans characteristically responded one way to such pictures, the association mechanism *might* provide a good explanation of how that response is retrieved. But, in fact, if Mr. A is presented with a picture of a friend, on two different occasions, he might very well respond differently. One time he might say, for example, "That's my friend Jack." Another time, he might say, "By George, what a good picture." So it would not be correct to say that the stimulus of the picture is connected to a particular response. And, of course, this example is in no way peculiar. Nor is the relation one-to-one in the other direction, from response to stimulus. That is, the response, "That's my friend Jack." might be made to the stimulus of the picture, or to some other picture of Jack, or to a painting of Jack, or even to the sight of Jack in the flesh. So we must admit, right from the start, that the connection between stimulus and response is going to be more complicated than a simple one-to-one connection. In other words, this extremely preliminary set of examples suffices to show us that we are not dealing with a jukebox-type retrieval mechanism. The associations formed between various stimuli and various responses are more complex than that.

It would be instructive to trace out some possible connections between a particular stimulus and its associated responses. Let us choose as a stimulus the spoken sentence, "And now we will sing the 'Star Spangled Banner.' " I believe that the stimulus-response theorist will be surprised at the results of this effort. (1). Suppose Mr. A is at a baseball game. As the announcer provides this stimulus over the public address system, Mr. A responds by singing the 'Star Spangled Banner', his face set in a noble expression as he sings it in his most operatic baritone. (2). Take another baseball game where the same stimulus is presented. Mr. A, who is, in fact, an opera singer, mutters to himself, "I really must preserve my voice for tonight's performance of *Rigoletto*." And that is his response. He does not sing the 'Star Spangled Banner'. Certainly this is a completely plausible response. Imagine that the last time Mr. A went to a ball game, his voice was hoarse as a result of his too-spirited singing. In fact, the reviews of the opera (it was opening night) all mentioned that he was a bit hoarse. Who could blame him for wanting to preserve his voice? But consider another response: (3). Mr. A, wishing to preserve his voice, sings softly, and leaves out the highest note. Surely he might do that. Or (4) suppose that, although he is worried about becoming hoarse, nevertheless he knows that the CBS reporter has a microphone aimed at him. He must sing the 'Star Spangled Banner', and in a hearty voice, or all his fans will be disappointed when they hear him on the evening news. So, when the announcer gives the stimulus, he sings in an especially lusty voice. (5). Suppose our opera-singing Mr. A is quite politically concerned. Disgusted with the executive branch of the government, and mindful of the dangers of dictatorship, he refuses to act like the other people in the crowd, who suddenly remind him of sheep, or of the crowds cheering Hitler in the middle 1930's. So, when the announcer provides the stimulus, he responds by saying, "Bah! Sheep.", perhaps sitting down with a disgusted wave of his hand. Alternatively (6) he may just look at all these sheep and, in his most stentorian tones, go, "baaaaaaa". Who would have thought that an imitation sheep bleat would have been an intelligible response to the stimulus, "And now we will sing the 'Star Spangled Banner' "? Remember, according to the stimulus-response theorist's account, Mr. A's bleat is a response which has been procured by the retrieval mechanism. That mechanism picks out (presumably) the sheep bleat trace. It can only do this in virtue of the fact that there is a prior association set up between this particular response and the stimulus under discussion. The association was one set up during some learning process. I mentioned this just now to remind us that all of these responses will have to be explained by appealing to some mechanism of association. Otherwise, the stimulus-response account of trace retrieval fails.

Before that moment of reckoning, however, it would be useful to continue with this list a bit longer. I beg the reader's indulgence; the purpose of the list is not to weary you. Mr. A might (7) instead of bleating at all the sheep, have shouted "Sieg heil!" at the top of his lungs, thus expressing his belief that our nation is rapidly becoming a police state. Or (8) in protest, he might have held up a big picture of Ho Chi Minh. (9). Alternatively, if Mr. A were a member of the John Birch Society, he might have held up a picture of Nixon and a picture of Ho Chi Minh. And while everyone was singing the 'Star Spangled Banner', Mr. A would be standing there with the two pictures, yelling, "Look! Two communists out to destroy our republic! One is already dead. Let's get the other one!" Suppose (10) Mr. A has a girlfriend in Duluth, Minnesota. One day two years before the baseball game in question, Mr. A took his girlfriend to a baseball game in Duluth. When the announcer in our example provides the stimulus, Mr. A thinks of that ballgame, and sighs, whispering, "Ahh, Duluth, Margaret."

The point should be clear by now. Human life is complex enough so that a given stimulus may be followed by virtually *any* response. If the stimulus-response theorist is going to stick by his guns, he will have to admit that *any* response may make its appearance at the ballgame. The example of the stimulus is in no way special. It was chosen without prior deliberation; and the examples of possible responses also were chosen without deliberation, in an almost offhand way. I maintain that all the examples are perfectly possible, and quite plausible. Is there anyone who would maintain that one or more of these examples could not happen? And, of course, the list of possible responses could be continued indefinitely. Thus, the mechanism of association will have to be one which has formed connections between a given stimulus and virtually every possible response. And, in the vast number of cases where the response must involve a trace, the stimulus must be associated with virtually every trace.

Before we examine the full import of this conclusion, we should investigate the process of association going in the other direction, from a particular response to the stimuli which may elicit it. Once again, let us call upon Mr. A to provide us with examples. This time, the response that Mr. A comes out with is the number and street address of the house he lives in (18 N. Eleventh St.). Here are some of the stimuli which, I stoutly maintain, could conceivably elicit that response from Mr. A. First, of course, we can imagine Mr. A giving someone his home address. But, surely, this is not a case of a *single* stimulus eliciting the response in question. Imagine: (1). Mr. A's best friend says, "What is your new address?" He answers, "18 N. Eleventh St.". (2). The friend says, "Did you move yet?" Mr. A nods and says, "18 N. Eleventh St." (3). The friend says "Where do you live now?" (4). The friend says, "In what locale are you hanging your hat these days, old buddy?" (5). The two friends make plans to have dinner together. Mr. A's friend is writing himself a reminder note. At the bottom of the note he writes, "5 p.m. – 300 Chestnut Lane." (That is Mr. A's old address.) So Mr. A shakes his head and says, "I've moved." His friend looks up in surprise, as Mr. A says, slowly and distinctly, "18 N. Eleventh St." (6). After Mr. A has corrected his friend's mistake, his friend starts writing the new address at the bottom of the note. But he gets stuck and says, "How's that again?" Mr. A responds. (7). Or his friend gets stuck and says, "18 what?" Mr. A responds. (8). In a new situation, a policeman stops Mr. A, who is driving rather recklessly. The policeman looks over Mr. A's driver's license. He says, "O.K., Mr. A, what's your present address?" (9). Or the policeman says, Harold (that's Mr. A's first name, printed on his license) where do you live?" (10). Or the policeman misreads the first name on the license, and says, "Harley, where do you live?" Mr. A, who thinks he can talk this cop out of a ticket, since the cop has addressed him in a friendly conversational way, does not feel like correcting the policeman. So he just says, in his meekest, law-abiding tones, "18 N. Eleventh St., sir." (11). Or, when the police pull him over, Mr. A doesn't get a chance to be friendly and sincere. The cop just barks out, "Name and address!" Mr. A, a friend of the police commissioner, mocks the policeman by barking his address in a voice which beautifully imitates the policeman. (12). Mr. A is in a strange city. As he walks around, he sees a building which vaguely reminds him of his own apartment building. Further, this building also has the number 18 on the front steps, though they are green numbers, rather than the red ones on Mr. A's own apartment house. Mr. A looks up and sees that this building is, by coincidence, on N. Eleventh St. In mild surprise and amusement, he exclaims, "18 N. Eleventh St., by George." (13). Mr. A walks by his old neighborhood. As he walks by his old house, he says, thinking affectionately of his happy years in apartment #3, "Ahh, 18 N. Eleventh St. How happy I was then." (14). Walking through the woods in British Columbia, Mr. A sees a squirrel. Thinking of his pet squirrel at home in Chicago, and wondering if his friend would feed the squirrel and take care of the house in his absence, Mr. A says out loud to himself, "I hope all is well at 18 N. Eleventh St." (15). Sitting at home watching tv one afternoon, Mr. A is exposed to a commercial jingle. Two girls, dressed in parachutist's garb, smile at the camera and sing, "Krispy Krunchee Shredded Wheat can't be beat!" Mr. A, a hypnotised smile on his face, whispers, "18 N. Eleventh St. can't be beat."

Once again, the point of the list should be clear by now. Just as a given stimulus can elicit virtually any response, a given response is quite able to appear after any number of stimuli of the most diverse sort. Once again the example was chosen with almost no forethought. Translating this result into the terms of stimulus-response theory, we see that the theorist is committed to something like the following account. Each time Mr. A says "18 N. Eleventh St.", we must suppose that this response is the result of trace retrieval. The retrieval mechanism picks out the trace of

Mr. A's address. This is, of course, a requirement. Otherwise, there would be no way of explaining how Mr. A got the address right. The trace didn't get picked out at random, however. Rather, according to stimulus-response theory, there existed an association, a connection, between the stimulus and its response. But, since we have seen that many disparate stimuli can elicit one response, and since we have seen that one stimulus may elicit many disparate responses, we must conclude that each stimulus is associated with many, many responses, and that each response (each trace) is associated with many different stimuli. In fact, we could make a stronger statement. It almost seems as if *every stimulus* is associated with *every response*. We have no reason to doubt this. Certainly the stimulus of the announcer at the ballgame elicited a rag-tag collection of responses. They don't really seem to belong in a class by themselves. (Except, of course, the class of 'responses to the announcer's stimulus'.) And the list of possible responses to that stimulus was terminated for reasons of space, not for lack of imagination. If the theorist balks at our statement that the stimulus must be associated with *every* trace, we must admit that we can offer no proof that such is the case. But, clearly, the stimulus-response theorist cannot deny that each stimulus is associated with an *indefinitely large* number of possible responses (traces). And, of course, an analogous situation holds vis-à-vis the associations between a particular response and the indefinitely large number of stimuli which may elicit it.

But what does this tell us about the proposed mechanism of association? For the retrieval mechanism to work, there must be an association between the stimulus and the particular trace which represents the response to that stimulus. Think of this association as a mechanical connection between the two. How does the retrieval mechanism come to pick out the right trace? Is it magic? No, says the theorist. The retrieval mechanism is purely automatic, a jukebox. It picks out the right trace because there is a connection between that trace and the stimulus. But this hypothetical mechanism of association fails to explain anything. It would serve well enough if there were an extremely limited possibility of response to a given stimulus. But the whole thrust of our examples is to show that there can be no such limit. The possibilities are completely open. Given a particular stimulus, almost any response may follow. And given a particular response, it may have been evoked by virtually any stimulus. So, in terms of retrieval, we are back to the same old problem. Given a stimulus, how can the retrieval mechanism pick put the right response? If the stimulus, "We will now sing the 'Star Spangled Banner' " can elicit almost any response, then how are we to explain the retrieval mechanism picking out the trace of the 'Star Spangled Banner?' (Rather than, e.g., a trace of *Deutschland Uber Alles*?) The association mechanism serves no purpose here. It drops out as an effective part of an explanation of trace retrieval. Since this association mechanism is vital to the stimulus-response account of memory, the failure of this mechanism brings with it the failure of the entire account.

But not wishing to dismiss a widely-held theory out of hand, we will not abandon our examination of stimulus-response explanations just yet. It should be noted that some theorists are aware that this type of simple association won't do the job. Accordingly, they have devised various attempts at getting around the problem. We will look at some of these.

Most stimulus-response theorists would have considered our account vastly oversimplified. They realize that each immediate stimulus may be followed by any number of different possible responses. But this concept of stimulus is too narrow, they might say. "Of course the announcer's words might be followed by any conceivable response. But the announcer's words are only a small part of the *total stimulus*. We must include other sounds as well. The roaring of the crowd, for example. And, further, all of the other sense organs are receiving stimuli. And we can't rule out the stimuli provided by internal states of the organism. Stomach pains, for example. All of this provides what is called the 'total stimulus.'

And the total stimulus, my philosopher friend, is *very* complicated." Skinner takes this view.

It is presumably not possible to show that behavior as a whole is a function of the stimulating environment as a whole. A relation between terms as complex as these does not easily submit to analysis and may perhaps never be demonstrated.[8]

But this move is really nothing more than an attempt to take refuge in complexity. The essential idea behind stimulus response is simple and clear. A physical stimulus brings about a physical response in a purely mechanistic way. However, when we look at an actual example, we see no such simple correspondence of stimulus to response. So the theorist asserts that the stimulus is so complex that, perhaps, we will never be able to show this correspondence. Nevertheless, it is there. Is this plausible? The idea here is, I suppose, that there actually *is* a one-one connection between stimulus and response. Now, however, we must include the fact that the room temperature is 72 rather than 71 degrees, as a part of the stimulus. And, presumably, the rippling of a particular muscle in the digestive tract as a part of the response. Taken in this way, there could never be anything but a one-one connection between stimulus and response. For no stimulus would ever be repeated and no response would ever occur twice. In the incredibly unlikely event that the same words were spoken by the same person in the same room in the same tone of voice in the same lighting conditions, etc. etc., the theorist could always fall back on a more precise measurement of temperature. "It was a different total stimulus because the ambient temperature was 71.83 degrees Fahrenheit, rather than 71.84 on the occasion of the previous stimulus. And, after all, such a temperature difference, though rather small, would definitely cause a difference in the rate of chemical reactions. And, after all, neurons fire as a result of chemical reactions, so . . ." This way around the thorny problem of retrieval will not do the job, however. For, even if this does preserve a one-one association between stimulus and response, it ignores another equally vital part of the theory. The process of association was supposed to be the stimulus-response account of *learning*. Learning is the process of connecting stimuli with responses, so that, in the future, a repeated stimulus will bring about a new response, or something of the sort. And if the total stimulus must be taken into account, there are no repeated stimuli. The concept of the 'total stimulus' contributes nothing to a stimulus-response account of memory.

The concept of 'total stimulus' runs into another problem, too. Suppose that, #1: on one occasion, Mr. A is presented with a total stimulus involving the announcer's words, "We will now sing the 'Star Spangled Banner.' " Suppose that, on this occasion, he does sing it. On another occasion #2: he is presented with a total stimulus involving the same words, but he responds by bleating. In order to explain this differing response, the theorist invokes the concept of the total stimulus. For example, it was raining on the last occasion, but it's sunny on this occasion. So far, so good. But let us take a third occasion #3. The accouncer says, "Our national anthem." And Mr. A sings the 'Star Spangled Banner.' Even though the stimulus (the announcer's words) is different from occasion #1, the response is the same as occasion #1. Of course the *total* response differs from the first occasion to the third. We can be sure that Mr. A's stomach muscles behaved differently on the first occasion as compared to the third, for example. But in both cases, the trace of the 'Star Spangled Banner' was recalled. Isn't that amazing! Considering how different the two stimuli are, on occasion one and occasion three, it is quite a coincidence that the two different total stimuli should call forth responses involving retrieval of the same trace. The theorist is not free to point out the similarities in the two total stimuli. Admittedly, Mr. A is at baseball games on both occasions. But then, he was at a baseball game on occasion number two just as well. And there the theorist felt it natural to say that the two occasions (i.e., one and two) were different (sunny and rainy days). So naturally the response would be quite different. Yet the theorist must ignore these (suddenly unimportant) differences when occasion three comes under consideration. Such a procedure is obviously illegitimate. It would, on this view, be a matter of *coincidence* if the trace of the 'Star Spangled Banner' were associated with these two very different stimuli (occasions one and three) and was not associated with the stimulus on occasion two. This coincidence would, of course, be as magical as the coincidental recall of the right book from a huge library.

And, of course, it is *not* a matter of coincidence that Mr. A sings the 'Star Spangled Banner' when the announcer calls it by name, and when he calls it our national anthem. Our national anthem is the 'Star Spangled Banner'. A stimulus-response theory which relied solely on the notion of a total stimulus to explain the fact that the 'Star Spangled Banner' trace is retrieved on occasions one and three, would be ridiculous. Clearly what is required is some sort of organization or classification of the stimuli. In other words,we need a mechanism which somehow puts together various stimuli. In this case, the mechanism would be one which brings about retrieval of the 'Star Spangled Banner' trace, in the two examples where the stimuli involve the words, "We will sing our national anthem", and "We will sing the 'Star Spangled Banner'." There are plenty of theories of this type.

But if the stimuli are to be organized, then there must be some mechanism which accomplishes this task. The mechanism must also bring about responses which are somehow related to the stimuli. The mere idea of the total stimulus would make it a matter of chance as to what response follows a given stimulus. Now, there are various ways in which these responses might be related to the stimuli which elicit them. Here is an example of how one author solves the problem of stimulus organization.

Let me now extend what I have said about individual learning to what would superficially seem to be much more complex phenomena. Those of interpersonal interchange. To do this, we have only to personify the experimenter as well as the learning subject, and to see the learning experiment as a small segment of an interchange between two persons. A, the experimenter, provides the stimulus. B, the subject, responds to the stimulus. And A follows B's response with a reinforcement . . .[9]

Bateson provides what he considers to be a 'general schema for an on-going interchange between persons who behave alternately'. This schema is supposed to apply to all interpersonal interchange involving two people. The behavior of the two people in question is supposed to be analysed into a sequence of stimulus, response, and reinforcement. He points out that the starting-point is arbitrary, and that each individual item of behavior (whatever that means) can be simultaneously regarded as a stimulus for the next bit of behavior which follows, as a response to the preceding bit of behavior, and as a reinforcement to the bit of behavior two places behind in the interchange. We will not here criticize the notion that behavior is analyzable into discrete segments. Right now we are interested in the way Bateson thinks that learning takes place. (His theory defines any interpersonal interchange as a learning interchange.) In other words, we want to know how the stimuli are organized in such a way that they bring about appropriate responses. We want to know how the two stimuli 'Sing the 'Star Spangled Banner,'' and 'Sing the national anthem,' both cause recall of the same trace. Bateson hypothesizes a situation involving an interchange of some arbitrary number of 'bits of behavior'. Concerning a particular item in the sequence, he says:

The formal truth (e.g., that item 26 is simultaneously a reinforcement, response and stimulus) may not represent the natural history of the relationship as it is perceived by the participants. They are busy putting their labels, imposing their Gestalten, on the items . . .[10]

Now we see how it comes about that in both cases (where the announcer says "We will sing the 'Star Spangled Banner'." and where he says "We will sing our national anthem,") the same trace is retrieved by Mr. A's retrieval mechanism. These two different stimuli are classed together by Mr. A himself. For example, he realizes that both announcements constitute requests. And he simply realizes that the 'Star Spangled Banner' is our national anthem. But, of course, this won't do as an account of trace retrieval. The trace theorist is stuck with the same old regress. If Mr. A classifies both stimuli as being requests, then he must know what a request is. He must remember what a request is. How does he do that? Does he have a memory mechanism? And so on. In other words: The memory mechanism postulated here requires that the stimuli and responses be organized. How does this happen? Bateson proposes that Mr. A does the organizing, thus allowing the memory mechanism to work. But we have pointed out that, in order for Mr. A to organize the stimuli, he must already remember all sorts of things. How does this happen?

So Bateson's suggestion leads to the infinite regress of memory mechanisms. Each mechanism stands in need of another mechanism to explain the workings of the first. Another way of describing the problem with Bateson's suggestion is to say that Mr. A himself is the homunculus that operates the machine. If Mr. A can "put labels on and impose Gestalten upon the items" of behavior, then why does he also need the stimulus response mechanism? If he remembers what a request is, what an anthem is, what the national anthem of the United States is, etc., *without* his memory mechanism, then why does he need the mechanism at all? It is interesting to note that the infinite regress and the homunculus are really two sides of the same coin. When, for example, we have uncovered a homunculus, we can always ask the mechanist, "And how does the homunculus remember? What sort of memory mechanism does *he* have?" And, given a trace mechanism, we can always ask the theorist, "How does the machine know which trace to retrieve? What is the name of the homunculus who operates the mechanism, and where is he?" Given an infinite regress of such mechanisms, we can always point out that, in the end, there will have to be a homunculus to operate at least one of them.

While Bateson's homunculus is operating right at the surface level, organizing the stimuli and responses directly, there are other variants of stimulus-response theory which have a hidden homunculus. All of them are involved in one way or another with the task of organizing stimuli, responses, or both. I am not suggesting that the proponents of these theories knew they were hiding the homunculus within their hypothetical mechanisms. In fact, we can be sure that the homunculus is hidden from them as well. Perhaps this blindness on the theorists' part comes about as the result of too great a fascination with complex mechanisms; or perhaps it is a stupefaction caused by the use of too much technical jargon. Speculation aside, here are a few examples of stimulus-response theories requiring the hidden homunculus:

H. Chandler Elliott examines a case where an infant has learned the word 'cat'. He says that the baby must have developed a cat-engram. According to Elliott, it comes about that by and by the baby recognizes that "people often couple this thing with the sound 'pussy'."

> . . . the word is always the same with only minor variations, compared to the great variety of distances, positions, and other aspects of the real cat. One might picture a word-stub (simple engram) penetrating and tapering out in the more intricate web of cat-engram, which it activates with accompanying emotional response . . .
> Direct connection between hearing-stub and speaking-stub of a word engram is minor. In later life, heard word seldom evokes spoken word; and so the hearing stub rather ramifies and loses itself in the great silent areas like the ribs of a leaf. For example, "Tell John" is not parroted, but evokes memories of John, where he is, how to get him, and a whole background of relevant detail leading on to planned behavior . . . The whole mind intervenes between recornizing and uttering the name.[11]

The homunculus is required to do several jobs in this account. For example, how does the baby "recognize that people often couple this thing with the sound, 'pussy' "? What mechanism figures out that the people are coupling anything with anything? Does this mechanism know what 'coupling' is? Does this mechanism perhaps have philosophical beliefs about how the people come to couple the sound with the thing? Does it, for example, think that this is a case of constant conjunction? These questions sound ridiculous; but they could, I believe, legitimately be asked if Elliott's account is accepted. Another place where the homunculus would come in is Elliott's account of the adult. What memory mechanism can choose "memories of John, where he is, how to get him, and a whole background of relevant detail"? Who is the homunculus who decides whether a particular detail is relevant? And even if all of these details are connected with the spoken stimulus 'Tell John', how does the mechanism choose which of these already-recalled details to act upon? What homunculus is it that, in Elliott's words, chooses the details to use in formulating 'planned behavior'? Who does the planning? No, we can see that the mechanism in this account needs several people to run it. And they all must have complete memories of their own.

Having seen that a stimulus-response type mechanism will require a homunculus to bring about organization of stimuli and responses, we will not elaborate any further. For reasons of space, this section will conclude with a presentation of a few selected passages from the works of various stimulus-response theorists. In these passages, we can see that the proposed mechanisms could only operate under the direction of a human intelligence. Here, for example, is a mechanism which, under the guise of 'pattern analysis' is required to recognize things, to know what is going on. The machine must be able to see and hear, taste and smell, etc.:

> In contrast to existential discriminations, based on the presence or absence of a stimulus, differential discrimination of this sort logically would seem to require the nervous system to analyze the temporal sequence or pattern, of stimulation.[12]

In a work we have already referred to, W. R. Russell often seems confused about whether his mechanism remembers by means of purely machine-like processes, or whether it possesses an intelligence which works the machine. He says:

> Man's brain is concerned with analyzing afferent impulses in relation to previous memories and their emotional responses, followed by a response.[13]

What 'relations' could a machine analyze here? What *are* the relations between afferent impulses (caused by stimuli) and their associated responses? We have already seen that the connections between possible stimuli and possible responses must be virtually open-ended. So here, Russell interposes a mechanism between the stimulus and the response. He never, of course, tells us what sort of mechanism it is. All we know is that it is the sort of mechanism which can analyze what is going on, and respond appropriately. It is a person.

In another place, W. R. Russell says that the essence of memory is the recognition of familiarity.[14] We supposedly have a comfortable feeling of familiarity and security when faced with something we have seen before.[15] He never says whether this feeling is conscious or not. From my own experience, I can say that the feeling would not usually be a conscious one. We will not discuss the notion of an 'unconscious feeling'. We can, however, ask the name of the homunculus who is supposed to recognize this familiarity. Whoever he is, this homunculus presumably possesses a memory of his own. A machine which reacts to familiar things won't be a *component part* of a memory machine. It must already remember. At any rate, Russell's homunculus will not only recognize things, it will also play a part in producing a response, in action. For example, this mechanism also plays a crucial role in speech. The reaction of familiarity takes place not only toward sensory stimuli, but also toward certain words. That is, the 'familiarity recognition mechanism' causes us to pick out the right word, when we are about to say something. It reacts toward the word which fits what we are trying to say.[16] We cannot deal with the host of philosophical issues which his suggestion raises. (E.g., does he think, then, that what we are going to say is already picked out? If so, then what mechanism picks it out in the first place? Further, does he think that our words are stored somewhere? Where?) But, for the purposes of our examination, we can ask, "How does the mechanism know which word to react to?" The mechanism must say to itself, "Ahh. That's the word we were looking for." The mechanism must already be a speaking, thinking, remembering homunculus. But Russell's homunculus has a further task. It not only speaks for us, but it directs our movements as well.

The execution of a skilled movement involves recognition of a familiar pattern combined with an emotional reaction which is disturbed by a performance which deviates from the familiar form.[17]

Now this 'familiarity recognition mechanism' has the job of telling us how to do and say the right thing. If, for example, we are about to insult someone, then this is the mechanism which chooses the word 'nincompoop', and, presumably, guides the movements of the vocal cords. Or, if we are about to attack someone, this is the mechanism which picks out the right 'action pattern' (throwing a stone, for example) and which guides us in the execution of this 'skilled movement'. But Russell says that all of this takes place after the decision is made to (e.g.) insult or attack someone. Speaking of this recognition function, he says:

This doesn't, of course, tell us how, at a higher level, the decision to throw a ball is taken.[18]

Russell sees that this function will require another mechanism. This will be the mechanism which makes decisions of policy. Once the 'recognizer' tells it, "Hey, we have been insulted." this 'high level decision mechanism' tells it, "Well, then, let's insult them right back. Er, who was it, by the way?" The 'recognizer' says, "It's your (our, my?) best friend, Hymie." Presumably, the decider would instruct the recognizer to pick out a fairly mild insult, or one which, if not mild, will at least be taken in fun. Although this is an admittedly comical way of explicating Russell's view, it is, I think, justified. The two mechanisms involved *would*, in fact, have to perform these functions. And, of course, the problem is that each 'mechanism' would already have to possess a complete memory of its own. Indeed, both the decider and the recognizer would have to possess personalities of their own. (A more effusive recognizer might have chosen the mild insult, "You silly boy, you." And a more irascible decider might have instructed the recognizer to pick out a really shattering insult.) So here are two complete homunculi hidden away in W. R. Russell's proposed mechanism. He even tells us where the decider lives – in the frontal lobe system of the brain.[19]

We conclude the section on stimulus-response memory theory with a few quotes. All of the proposed mechanisms require a homunculus to operate them:

Recognition, if it has a nervous basis at all, must be a result of comparing a present stimulus with traces left in the nervous system by past stimuli or sets of stimuli.[20]

Some stimulus-response memory theorists assume that the TBRU is coded by category of the elements of which they are composed . . .[21]

. . . a new experience is somehow immediately classified together with records of former similar experience so that judgment of differences and similarities is possible.[22]

When an item is presented a 'sense of familiarity' is aroused. When tested, the subject bases his response on a partially faded trace: low degrees suffice for recognition, while a greater degree of familiarity is needed for recall.[23]

If memory is like a junk box or a filing system, then the process of learning to retrieve could be accomplished by getting the item separated from the rest of the contents of memory in some way, or by getting the contents of memory organized in some systematic way so the subject knows where to look for things. There is another way of thinking about memory that may be more realistic . . . Another possibility is that memory structures or engrams are functional as well as structural features of the mind. On this view, a stored memory structure becomes active when an appropriate signal is received – the engram may be thought of as waiting for its number to be announced before coming forward. If memory storage involves establishing engrams, then the question of retrieval is the question of whether the engram becomes active when the stimulus is presented on the test. And if it does not with sufficient reliability, then the subject has to set or tune the engram more efficiently so that it will be activated reliably by the presentation of the stimulus.[24]

Stimulus-response theories of memory do not, then, offer an advantage over other trace-type theories of memory. Just as the other theories failed on the problem of retrieval, stimulus-response theories fail on the problem of association. And the mechanism of association is just the prerequisite to successful retrieval. Thus, stimulus-response theories fail on the problem of 'pre-retrieval', if you will. That is, they do not solve the retrieval problem at all; they merely move it back one step. The homunculus is hidden in the association mechanism, rather than in the retrieval mechanism.

INFORMATION PROCESSING COMPUTER THEORIES OF MEMORY

Having concluded our examination of stimulus-response theories of memory, we will now turn to the view usually referred to as 'information processing'. Our reasons for dealing with it at all are the same reasons which caused us to take up our examination of stimulus-response theory. It is not, for example, that information processing theories are significantly different from other mechanistic theories of memory. Like any other plausible mechanistic theory of memory, information processing is a trace theory. As such, it falls prey to the criticisms which have been brought out in the course of this study.

But, as a result of the birth of computer science, much excitement has been generated in and around our scientific community. Here, at last, is the mechanism which is capable of possessing intelligence. Here, finally, say the theorists, we have machines which can do almost anything people can do. Or, if this is at present an exaggeration, certainly someday there will be computers which speak, read, write, think and remember. Or so the scientists say. Computers have even captured the popular imagination. There are books, indeed there are movies which feature computers of human or even superhuman intelligence. Only backward-looking romantic, starry-eyed humanists could fail to see that computers are real examples of intelligent mechanisms. People, too, are supposed to be ultra-sophisticated computers. Some of the operations our computer-brains perform are the processes of memory. What could be a more reasonable assumption?

Unfortunately for these theorists, however, we have seen that it is *not* reasonable to explain the phenomena of human memory by assuming the existence of memory mechanisms. For, whatever their design, all such mechanisms will require either a homunculus to run them, or an infinite series of mechanisms. And this truth holds whether we consider computer trace theories, or stimulus-response trace theories, or any other trace theories. The criticisms we adduced against such theories do not depend on the particular design of the mechanism. Rather, the theories are open to the criticisms because they are mechanistic trace theories. And so, information processing theories are ruled out along with the others. In the final section of this book, we will see that the mistakes which are inherent in trace theory have a philosophical basis. But before dealing with that, we will look at information processing computer theories just long enough to see that they do not differ significantly from the other theories we have examined.

As a preliminary point, it should be noted that, like stimulus-response theory, the computer view is often applied to phenomena other than just those of memory. But, insofar as the proponents of views of this type do deal with memory, they do so in terms of memory traces. And, of course, this is not accidental. As we have noted a number of times, Wolfgang Köhler was right when he said, ". . . no theory will be acceptable which fails to assume the existence of some trace."[25]

Central to computer processing theories of memory is the concept of *information*. The source of this concept is supposed to be mathematical theories which describe the operations of actual computers, i.e., ones which have been manufactured out of electronic and mechanical components. These machines contain varying numbers of switch-like components. If we feed an electrical pulse into such a machine, the current flows through these switches. And the flow of current within the machine is dependent upon the state of these switches. For example, if some of the switches are in the 'on' position, current will flow through them. If the switches are in the 'off' position, the current must flow some other way. The exact description of such current flow is a matter for engineers to predict and determine; it need not concern us here. But the various states of the machine are referred to as *states of information*. The more possible paths through which the current may flow, the greater the 'uncertainty' of the machine. As more and more pathways are ruled out, the smaller the amount of uncertainty in the machine. The pulses which cause these pathways to be blocked off are defined as *information*. We can see that this is a highly technical concept. Information, as a technical term, is completely dependent upon the state of the switches in the system.[26] The following passages characterize and define this technical concept, as well as provide an example of information flow within such a machine:

Information (as defined by Shannon) is merely a measure of how much uncertainty has been removed by receipt of a message.[27]
Definition of 1 bit: the amount of information necessary to resolve two equally likely alternatives.[28]

On page 6 the authors present what they refer to as an 'admittedly elementary example of a guess-the-numbers game". Instead of punching numbered buttons on the machine, we are to imagine that we are asking questions. This will serve as a sort of bridge between information theory as applied to machines, and information theory as applied to people. In their example, we are to guess a number from 1–8. We can ask questions like, "Is the designated number in the group 1–4?" If the answer is, for example, 'No,' then we know that the designated number is in the group 5–8. So then we can ask, 'Is the designated number either 7 or 8?' If the answer is, for example, 'No,' then we know it is either 5 or 6. Our last question will, of course, determine which number was originally designated. It is essential to this example that our questions could be answered by a "yes" or a "no". This would allow us to substitute buttons or switches on the machine for spoken questions. Here is their example:

Let us designate the 8 equally probably states of the system by the numbers 1–8. Assume that you are to ask only binary (yes or no) questions in an attempt to determine the state of the system. It is a simple matter to discover that the minimum number of questions here is three. In this simple illustration we have introduced the basic concepts from which the quantitative definition of information can be formulated. Namely: the number of equally probable states of a system (in this case 8) the number of alternatives resolved by each question (2 because of the binary nature of the question) and the minimum number of questions necessary to determine the state of the system. (3 in this case).[29]

Now, clearly, there is no philosophical objection to this basic concept of computer theory. It is extremely useful to those who are involved in the design and employment of computers. 'Information' here is a quantifiable concept. Its use allows us, presumably, to describe (e.g.) the storage capacity of a computer in terms of the number of possible combinations of switch pathways built into the machine. Each possible combination is a *state* of the computer. It would be a mistake, however, to confuse this technical concept with our non-technical word, 'information'. While 'information' in the ordinary sense is, for example, subject to quantification, it is not quantifiable in a precise manner. We can say that *A* is a better scout than *B*, because *A* always brings back more information than *B*. But we can't, for example, say that *A* brings back 36% more information than *B*. That would be a ridiculous thing to say. It is clear that information in the technical sense is different from what we would usually take information to be.

But there is a much more important reason for keeping in mind the fact that computer theory uses the word 'information' in a highly technical sense. It is, by way of illustration, true that computers can possess and even exchange information, in the technical sense of the word. But, of course, it does not follow from this that computers can possess or exchange information in any other sense. It does not follow that computers can talk to one another, write each other letters, exchange information with people, or anything else. When we say that a certain computer has been programmed to contain some information, we mean that its internal switches have been disposed in some particular way which will cause it to (e.g.) print out a list of telephone numbers when we push a certain sequence of buttons. This does not imply that the computer has any information (non-technical sense), that the computer is well-informed on a certain subject, that a computer knows anything, thinks anything, etc. It would, however, be reasonable to conclude that someone who exchanges information with someone else (non-technical sense) *can* be well-informed on a certain subject, can (and does) know something, can think about things.

It would be an egregious error to confuse this technical sense with the non-technical word. An example may be of use here. Suppose a credit agency's computer sends out a threatening letter to a client whose bill is overdue. Suppose the letter is actually addressed to the client's computer (e.g., General Motors' computer). The letter might even be in the form of a computer punch-card. It would be a joke (maybe not a funny one) to say that the first computer was a threatening sort, or was intolerant of dead-beat clients, or something similar. Yet it might not be a mistake to say that the two computers exchanged some information (technical sense). We can imagine a computer theorist saying something like: "Well, nevertheless the two computers exchanged some sort of information. And if they can do this with today's technology, maybe someday they will be able to threaten each other, and so on." Let us make an analogy. Suppose that we have a technical term, 'desire'. Desire is the attractiveness that magnets show toward pieces of iron. Now we admit that this is a technical sense of the word. We can, for example, quantify magnetic desire in very precise terms, whereas we can only quantify desire (non-technical sense) in very imprecise ways. (E.g., "His desire for an ice cream cone was outweighed by his desire to lose weight." But we would never say seriously, "His desire for an ice cream cone was 17.3% weaker than his desire to lose weight.") Nevertheless, our magnetic theorist says, "The magnet does have desires, in *some* sense of the word. Maybe someday, when technology builds more sophisticated magnets, we shall have magnets that desire things in other senses as well." Please note, it does not matter if, someday, scientists build magnets that attract female humans. This will not show that magnets desire anything in any non-technical sense. A magnet in those futuristic days could no more (e.g.) be in love with Racquel Welch than today's magnets can be in love with chunks of iron. The fact that magnets *desire* (technical sense) iron does not show that magnets desire (non-technical sense) anything at all. It does not matter that two computers can exchange *information* (technical sense). This could never be used as grounds for saying that computers can exchange information in a non-technical sense. Further, it could never be used as grounds for saying that computers perform any other human sort of actions. For example, the credit agency's computer does not threaten the General Motors computer, even when the information exchanged results in a threatening letter being sent. On a completely different topic, Wittgenstein makes an interesting observation:

> Why can't my right hand give my left hand money? – My right hand can put it into my left hand. My right hand can write a deed of gift and my left hand a receipt. – But the further practical consequences would not be those of a gift. When the left hand has taken the money from the right, etc., we shall ask: "Well, and what of it?"[30]

The preceding three pages are not offered as a proof that computers can't think. Rather, they are intended as a reminder not to confuse the technical term 'information' with the ordinary word.

It is important to prevent this confusion. Otherwise, we might become embroiled in the following misguided chain of reasoning: (1) It is an undisputed fact that computers can process information. (2) When a computer receives some information, it has learned something. (After all, what is learning, if not the acquisition of information?) (3) Computer theory can be applied to learning and memory phenomena in people. That is, since computers can learn things, and can remember what they have learned, maybe people are computers too. Thus is born the information processing computer theory of human memory.

What we are pointing out here is that this line of reasoning involves a confusion of technical terms with ordinary words. In view of this possible confusion, it seems reasonable to suggest that the technical term 'information' be replaced by some other term, one which would be less liable to mislead. And it would be wrong to think that this confusion is an unlikely possibility. In fact, I think that some proponents of computer theory do fall into this trap. Gregory Bateson, for example, seems to think that he has solved the problem of whether or not computers can learn.

> Learning, as the word is used by psychologists, denotes the receipt of a meta-bit. i.e., a piece of information which will change the subject's response to some bit . . .
> this is *first order learning*. (Bateson's italics). By including this simplest phenomenon, i.e., the receipt of a single bit, within the spectrum of learning, the question as to whether a computer is or is not learning when it receives appropriate input, is answered out of hand. This is learning of the simplest order.[31]

Bateson is certainly not alone in his confusion. Several pages back, we saw a definition of the technical concept of 'information'. It was found on page 6 of *Information Storage and Neural Control*. On the very same page we find the following passage. The authors seem entirely unaware that they are confusing the technical term with the ordinary word. Perhaps they think that their technical definition also defines the ordinary sense of 'information'. At any rate, here is what they say:

Information theory does not deal with the importance of the information in a message. For example, the information in the message, 'The baby is a boy', is one bit, independent of whether you are the father. This comment is made in order to emphasize the fact that information theory does not deal with the subjective value of information, which falls more properly into the domain of semantics, but rather with objective measures of information.[32]

It would be fair, I think, to call this sloppy thinking. Certainly the authors' remarks have not been carefully thought out. A tremendously important philosophical and, I think, moral issue (whether or not people are machines) is to be answered in an offhand way, 'out of hand', by so-called men of science who can't even recognize when they are confusing a stark technical term with a complex and subtle human concept. This kind of offhand speculation leads to some amazing ways of looking at things. Do Fields and Abbott (the editors of *Information Storage and Neural Control*) assume that there is to be a science of semantics, which is to provide a description of the 'subjective value of information'? What are they talking about?

The confused idea that computers learn and know things, results in a tendency to personify them. If people are computers, then computers can someday become people. As computers become more complex and sophisticated, they will approach and eventually surpass humans in many areas of endeavor. Culbertson[33] for example, thinks that machines may someday excel in writing poetry.[34] Not only that, but he feels that someday, when the human computer (the brain) is improved in its operations, we will be better able to calculate and arrive at answers to ethical problems.[35] So machines hold the answers to crucial problems of right and wrong. Solomon-like computers are to show us the way to an exact science of ethics. This ludicrous vision, based on confusion, is, I believe, appropriate to a civilization which (as has been pointed out by many) deifies technology. We can imagine people someday consulting the High Priest and Head Programmer, who in turn consults the Great Computer. "Should we execute this criminal, oh Great One, or should we set him free?" This is no joke. Culbertson apparently thinks that someday, machines will tell us what is right and wrong.

But, if computers are someday to possess information, knowledge, wisdom, thought, consciousness, etc., then, for example, they must know things. And this means that they must remember things. And, unless these computers are to work by magical incantations, they must possess memory mechanisms. And that means that they must possess trace mechanisms. And that means that they must possess retrieval mechanisms. And retrieval mechanisms, as we have seen, require someone to run them.

Computer theory tells us how the traces are to be stored. The switches that we mentioned earlier form the basis of the memory storage system. To take a very simple example, a black-and-white image of the Mona Lisa can be created out of a pattern of black or white squares. If the squares are arranged in a convenient grid pattern, then they can easily be assigned spatial coordinates. Each spatial coordinate will be either black or white. And this would correspond to a switch being either in the 'on' or 'off' position. Thus, inside the computer, a particular group of switch circuits could represent a stored image of the Mona Lisa. This example is, of course, very crude, but it will serve as an illustration of the way in which a computer can store memory traces.

It has been suggested that the brain, too, contains structures which are analogous to so many on-off switches. Naturally, this suggestion was taken up by proponents of the human information processing computer theory. The structures which take the place of computer switch-circuits are none other than the neurons of the brain and entire central nervous system.

In modern times experimentation has supplemented theoretical thinking, and the search for mechanisms capable of explaining how sensory patterns are processed and stored in the brain has become a major aspect in physiological and psychological research. The tremendous development of digital computers has influenced theory-making in the biological and psychological fields, and has also allowed the possibility of checking theoretical models in actual operation by the simulation of hypothesized neuronal networks or by the use of programs that simply implement the logical operations believed to take place in the process being studied.[36]

The neuron has the binary decision to fire or not to fire, which is based upon the strengths and characteristics of its inputs and the present state of the neuron.[37]

To psychology, however defined, specification of the net* would contribute all that could be achieved in that field – even if the analysis were pushed to ultimate psychic units or 'psychons', for a psychon can be no less than the activity of a single neuron. Since that activity is inherently propositional, all psychic events have an intentional, or 'semiotic' character. The 'all or none' law of these activities, and the conformity of their relations to those of the logic of propositions, insure that the relations of psychons are those of the 2-valued logic of propositions. Thus in psychology, introspective, behavioristic or physiological, the fundamental relations are those of 2-valued logic.[38]

The retrieval mechanisms in information processing computer theories bear a striking resemblance to the retrieval mechanisms found in other types of trace theory. Some theorists, for instance, see themselves as presenting a combination stimulus-response and computer theory of memory.[39] In such theories, it is the computer which is supposed to perform the task of choosing the right response – the right trace. As we saw before, this mechanism would already have to possess a memory. Calling this mechanism a 'computer' doesn't get us around the problem of retrieval.

One very popular way to make sure that the retrieval mechanism knows how to find the right trace is to provide a mechanism which organizes the information to be stored. This process is the same as the process of stimulus organization, which, in stimulus-response theories, is supposed to set up an association between stimulus and response. In these computer theories, the *information* is analogous to the *sensory stimuli* of stimulus-response theories. But, just as in stimulus-response theories, this type of mechanism explains nothing. First, the organizing mechanism already requires a homunculus to recognize the stimulus (or here, the 'information') for what it is. For example, a television commercial would be recognized as

* By 'net', the author means a network of neurons which are interconnected, like the electrical circuits in a computer.

being 'information' which is separate from the quiz show it interrupts. To do this, the machine must remember or know what a quiz show is. After all, the visual characteristics of the quiz show itself can be quite varied. Think, for example, of a quiz show which has a few film clips thrown in as part of the show. How does the 'information organizer' mechanism know that these film clips aren't commercials? And we must assume that it does know this; when Mr. A is asked who was on the quiz show, he does not say that one of the contestants was Arthur Godfrey, who was selling Plymouths. So the 'information organizer' must remember that (e.g.) film clips are sometimes presented in the middle of quiz shows. And thus it must remember what a film clip is, what film is, what cameras are, etc. If it remembers by means of a further mechanism then we are involved in the familiar regress. And if it just remembers, without a further mechanism, then it is a homunculus. As the reader has noticed, this is the same line of reasoning which leads us to reject stimulus-response theories.

The second problem with 'information organizer' mechanisms is that they do not tell us how this organized information is to be used. If the stored input is to be utilized, then there must be a further mechanism which consults the information and decides how to use it. This 'output organizer' is analogous to the 'response organizer' in stimulus-response theories. And the 'output organizer' also requires a homunculus who knows (e.g.) that it is not appropriate output to dance in the aisle at a formal classical music concert. Such a 'mechanism' would be required to know and remember what the word 'music' means. In computer theories, this 'output organizer' is often a little black box called an 'executive programmer' or a 'high-level decision maker'. But the use of such jargon does not change the need for the homunculus inside the black box.

First, we will look at some theories which require the input information-organizing homunculus. In a move which we have seen employed by stimulus response theorists, some computer theorists put the homunculus right out front. They leave it to the subject himself to organize the information.

But beyond this, there is obviously a third order learning called the acquisition of 'test wisdom', or 'set learning'. Here the subject learns that he is to be on the lookout for sequences of a certain sort in his universe which include both external events and his own behaviors . . . I shall speak of this as third order learning, referring to those changes whereby the subject who encounters and solves repeated problems of a certain sort comes to expect his universe to be structured in ways related to the formal structuring of these previous problems.
That is, each person is receiving bits of information, and these bits are already falling into place as yes and no answers to questions of which the person already has understanding. But the second order learning must also be occurring. i.e., he must be changing his identification and understanding of the questions to which the bits are answers. And third order learning must also be going on, namely he must be learning the characteristic patterns of contingency in this relationship.[40]

The authors here seem to have forgotten that they are offering mechanistic theories. Instead of giving us a picture of humans as machines, they seem to be saying that humans are computer programmers, who happen to possess portable computers. Thus they attempt to tell us how the machine works, forgetting about the human programmer. But, of course, their original task is to explain how the human programmer works.

In general, the capacity of this more permanent storage system is so large that information that is stored there must be organized in an efficient manner if it is ever to be retrieved. Then finally, when it is necessary to retrieve information from memory, decision rules must be used, both to decide exactly how to get access to the desired information and then to decide exactly what response should be made to the information that has been retrieved.[41]

Many of the models in this book have borrowed from the concepts of the information processing field. We talk about the encoding of information . . . All this comes naturally from the various sciences that have developed around the technology of computers. Yet there is increasing evidence that subjects in fact, attempt to organize and make sense out of all material presented to them, even in tasks where the experimenter feels this to be difficult or impossible. It is clear that long-term memory structures depend heavily upon the organization of the material contained within it, and it is quite clear that linguistic performance depends upon the application of rules to the syntactic and semantic structures of language, rather than simple associations among previously encountered items . . . If we observe subjects who are performing experiments in memory, we cannot escape the conclusion that they are struggling with the material to be learned, not by attempting to memorize the items by rote, but by their attempts to find some rule or structure which they can apply to the material. Perhaps the most encouraging aspect of these speculations is the increasing realization that we must eventually come to understand the strategies and rules used by subjects if we are to eventually understand human memory.[42]

Another notion we have seen before is the idea of a 'key', or catalogue. The traces are all represented in this catalogue. The right trace is picked out by a mechanistic process of matching. But, as we saw before, this just pushes the problem back one step. How does the mechanism know which is the right card in the catalogue? Finding the right card or key is just as much a problem as finding the right trace. In computer theory, it is fashionable to refer to this catalogue, this set of ancillary traces, as a 'neural dictionary' or list of attributes, or a code by means of which all of the input information is organized. This ploy works no better here than it did in the other trace theories we looked at. These retrieval mechanisms all need a homunculus to pick the right key, to read the cards in the catalogue, to read the dictionary, or to know which piece of information to decode. Yet we find this suggestion (that the information or traces are organized in some system) over and over again, as if it constituted a step toward solving the problem of retrieval.

Input of appropriate kind and strength crosses the individual boundary and is transduced into the proper form for nervous transmission. If a language or code is involved, it is translated by the decoder and classified by the perceiver in terms of a perceptual schema which represents the world as the individual has experienced it. Reference may be made to stored memories. There may be some recoding or other preparation of all or part of the input for storage in the memory.[43]

For instance, sufficient information to output (vocalize) words must be stored as part of the permanent long-term store. Show an individual a pictorial representation of a common object and he can name it. These 'output subroutines' are acquired early in life and must result in something like the formation of a 'neural dictionary'.[44]

Just as a computer stores content and address, so it may be that the contents (*what* is stored) of this neural dictionary are the output subroutines necessary to articulate words, whereas the locations (*where* it is stored) are specified in terms of semantic information which functions as retrieval cues.[45]

In the process of trying to solve specific problems, computer theorists lose sight of the fact that they are supposed to be analyzing the phenomenon of memory into machine-type operations. Here, for example, the author seems oblivious to the fact that his theory requires the computer to *remember* what is stored, and where it stored it:

In computers an item of information is stored in and retrieved from locations in the core memory: i.e., to retrieve a given item, one has to remember where it is stored . . . It seems to be more economic to suggest that the basic structure of the memory system used by the brain is not addressed by location, but by content.[46]

Another suggestion from these experiments is that the short-term memory mechanism involves active working processes of input coding and programming.[47]

In the following passage (a rather long one) we find a description of a memory mechanism in considerable detail. The opportunity of seeing the authors' overview of what is needed in a complete memory mechanism will have to serve as justification for presenting such a long passage. In this passage, we find various types of homunculi. We find, for example, that input information must first be recognized. This is to be accomplished by the same sort of *stimulus analyzer* that we encountered in stimulus-response theories. The authors do, in fact, refer to the mechanism as a stimulus analyzer. We also find an interpreter mechanism, which decides what to do with the information that has been recognized. Then, on the other hand, the authors slip into a way of speaking which makes it seem as if the individual himself must play a part in dealing with the input.

There is a good deal of agreement about the nature of the psychological processes that transform a physical image into a meaningful psychological form . . . First, a visual or auditory signal is transformed into a sensory representation by the appropriate sense organs . . . Our primary impression of a signal is of its meaning, rather than of its physical characteristics. There is a good deal of evidence that this identification has already been performed by the time the signal enters short-term memory, for the immediate memory of an item often is more related to its psychological encoding than to its actual, physical form. Thus, somewhere between the sensory storage system and short-term memory, the sensory image has its critical features analyzed, identified and interpreted.[48]

By perception, we include those processes involved in the initial transduction of the physical signal into some sensory image, the extraction of relevant features from the sensory image, and the identification of that list of features with a previously learned structure. By memory we mean the processes that act to retain the material that was sent to it from the perceptual system.[49]

The authors never trouble themselves by asking "How is a machine supposed to decide what features are relevant?"

. . . we need to specify the retrieval cues that subjects use when trying to answer questions about the material that the experimenter asks of them.[50]
Physical inputs are stored in their sensory form by the sensory information store, while critical features are extracted from each item and placed in the appropriate perceptual vectors. The vectors of perceptual features are transformed into vectors of memory attributes by the naming system. The type of response is based on the question asked of the decision process, the set of possible response alternatives, and the attributes remaining in each memory vector, either temporarily (as short-term memory attributes) or more permanently (as long-term memory attributes).[51]

Notice that all of these processes are supposed to take place as preparation for the memory process. But, of course, these preliminary processes already require homunculi with complete memories.

The sensory image system transforms the physical signal into a sensory image, . . . the perceptual system attempts to extract sufficient characteristics from it so that it can be identified unambiguously.
We assume that we can represent information by means of a list of its features or attributes. That is, visual material might be represented by a list of its spatial properties, spoken speech by a list of its phonemes or distinctive features, and permanently stored material by a list of its relationships with other material.[52]

The problem of identifying the physical signal can be considered that of attaching a previously learned name to each particular acoustical or visual input. The simplest way of doing this is to use a dictionary that matches physical features with psychological names. We assume that the naming usually takes place in a rush, with the perception of the physical signal being incomplete and noisy, and with several things competing for attention simultaneously.[53]
In this chapter we do not consider the details of the way by which features might be extracted from incoming signals. Rather, we assume that some sort of stimulus-analyzing mechanism extracts generalized features and presents them to the naming system. This process continues either until the naming system manages to identify the signal unambiguously, or until the image of the signal has disappeared from the sensory system. Thus, we can characterize the processing of the amount of incoming sensory information as something of a race between the amount of time that the information is available in the sensory memory, and the amount of time needed to acquire sufficient number of features so that the new material may be properly perceived and encoded.[54]

Here again we find the concept of a 'dictionary' or 'catalogue':

This scheme for both recognition and recall assumes several things about memory. For one, it is necessary to have a content-addressable storage system. That is, given a list of attributes, we can find them in storage without requiring any lengthy search process.[55]
The job of the naming process is to transform the set of features extracted by the perceptual mechanism into a form meaningful to the memory system. This task is accomplished by matching, as best it can, the perceptual features extracted by the pattern recognizer against the features belonging to each set of possible stimuli it has been told to expect.
Thus, the naming system incorporates a dictionary that gives the rules for transforming each particular perceptual vector into its corresponding memory vector . . . If the features present in the perceptual vector match unambiguously with only one dictionary vector, there is no problem in determining the proper name. However, if several dictionary vectors match the perceptual vector equally well, the subject has a choice of strategies to follow.[56]

The following diagram is taken from an article in *Models of Human Memory*. It is a diagram of what the author calls the 'Logogen Model'. It is interesting for two reasons. First, it provides a graphic representation of the homunculi that the other authors have described in words. But second, it highlights the use of jargon and pseudo-scientific techniques which lie behind virtually all of the theories we have examined. At any rate, here is the 'Logogen Model', replete with not one, but five separate homunculi.

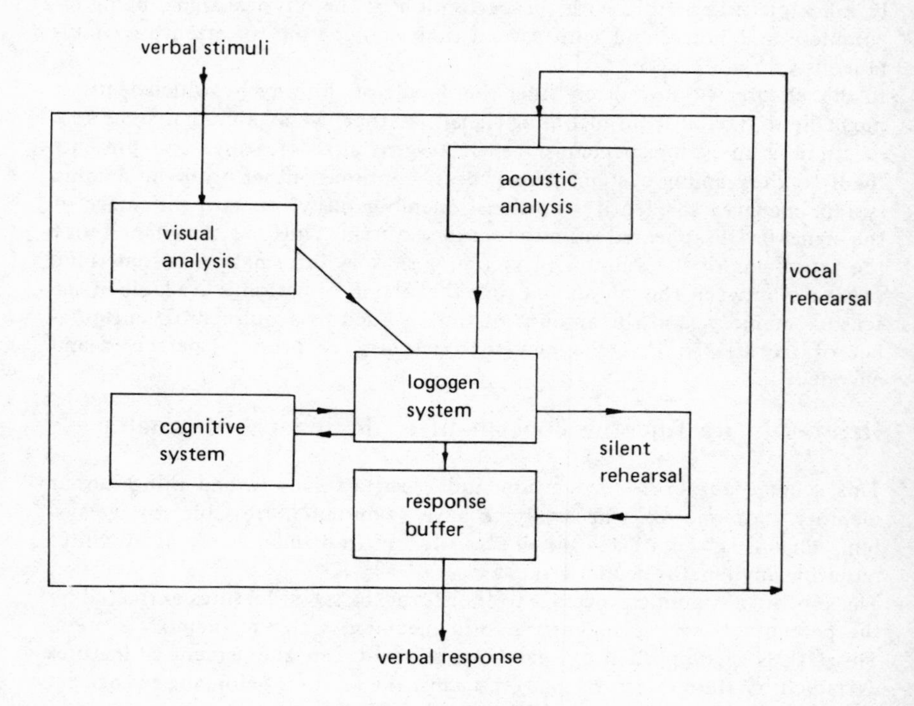

Even if we grant to the computer theorist that some mechanism might accomplish the (impossible) task of organizing or encoding the input information, we are no closer to an explanation of memory. The problem of input is linked, naturally, to the equally troublesome problem of output. Just as stimulus-response theories require a response-decider homunculus, so, too, do computer theories. Here, however, the term 'output executive' might be more appropriate. But the function is the same in either case. The executive decision maker must decide what to do in response to the analyzed input. And, as we saw in our examination of stimulus-response theory, any particular input can be followed by any one of an indefinitely large number of possible outputs. There is no simple connection between a given input and any particular output. Thus, we require a decider mechanism which not only must possess a complete memory (the regress), but which also possesses a complete personality. It must decide, for example, whether the boss' insult is to be taken meekly, or whether the individual will punch the boss in the nose and quit his job. The computer theorist's explanation of personal courage would then amount to something like this: "Mr. A is courageous because his decider mechanism chooses outputs that we call 'brave'." And we can ask, "How does the decider mechanism know what 'bravery' is?" We conclude the second part of the book with passages which refer to this 'output organizer', or 'executive program mechanism'.

A given set of inputs may elicit two or more alternative outputs. The decider selects the one that is put into action. Each of the subsystems of a system is also a system at its own level, and must make its own decisions, as well as carry out other critical functions . . .
The individual has a central decision-making subsystem which determines output for the whole system.[57]
Each one of a person's subsystems may participate in the preparation of the output . . .
On the output side, a decision is made from among the alternate possible outputs; encoding for external transmission is carried on, and the nervous message is transduced into physical response, through either the speech mechanism, or other musculature.[58]

This concludes the second portion of our examination. The mechanistic theories we have examined lead us down a blind alley. This impasse was reached by taking these theories seriously. When we attempt to follow an example through the steps required by the theory, we find ourselves in need of someone (the homunculus) to do the remembering. The mechanisms proposed therefore do not merely fail to give a *complete* account of memory. They do not even represent a partial step toward providing a scientific, causal or mechanistic account of memory. Apparently the theorists whose writings we have examined, fail to think their examples through to the end. Typical of this failure is Norman and Rumelhart's remark, "In this chapter we do not consider the details of the way by which features might be extracted from the incoming signals." Had they considered these 'details', perhaps they would have seen that only a homunculus (or a human being!) could do the job. This lack of perspicacity on the part of memory theorists is extremely puzzling. Certainly they are misguided in their theorizing rather than lacking in intelligence. Yet they all seem to end up in the same boat. In the final section of this study, we will try to uncover the unsound basis of mechanistic theories. As I hinted before, I believe that trace theories of memory are essentially philosophical, and not scientific, in origin. The misguided philosophical views from which trace theory arises are examined in the next section.

NOTES

[1] From pp. 8–9 of *The Behavior of Organisms*, by B. F. Skinner. Appleton-Century-Crofts, New York, 1936, 457 pp.

[2] *The Behavior of Organisms, op. cit.*, p. 433.

[3] P. 13 of *Brain Memory Learning*, by W. R. Russell, Oxford Press, 1959.

[4] See, for example, pp. 83–4 and 100 of *The Mammalian Cerebral Cortex* by B. Delisle Burns, *op. cit.*, or see p. 241 of *The Shape of Intelligence* by H. Chandler Elliott, *op. cit.*

[5] P. 257 of *Models of Human Memory, op. cit.*, from the article, 'How associations are memorized', by James G. Greeno.

[6] P. 259, *ibid.* This passage was quoted earlier.

[7] See pp. 122–131 of *The Task of Gestalt Psychology, op. cit.*

[8] *The Behavior of Organisms, op. cit.*, pp. 8–9.

[9] *Information Storage and Neural Control, op. cit.*, from the article, 'Patterns of human behavior', by Gregory Bateson, p. 175.

[10] *Ibid.*, p. 176.

[11] *The Shape of Intelligence, op. cit.*, p. 241.

[12] *Information Storage and Neural Control, op. cit.*, from the article by E. Roy John, 'Neural mechanisms of decision making', p. 243.

[13] W. R. Russell, *Brain Memory Learning, op. cit.*, p. 7. See also p. 14.

[14] P. 35, *ibid.*

[15] Also p. 35, *ibid.*

[16] P. 38, *ibid.*
[17] P. 24, *ibid.*
[18] P. 24, *ibid.*
[19] P. 131, *ibid.*
[20] *The Science of Mind and Brain, op. cit.*, p. 40.
[21] *The Biology of Memory, op. cit.*, From the article by Arthur W. Melton, 'Short and long-term postperceptual memory', p. 5.
[22] Dr. W. Penfield, quoted by T. A. Harris, in *I'm O.K. – You're O.K., op. cit.*, p. 10.
[23] *Models of Human Memory, op. cit.*, from the article 'Models for free recall and recognition', by Walter Kintsch, p. 333.
[24] *Models of Human Memory, op. cit.*, from the article 'How associations are memorized', by James G. Greeno, p. 261.
[25] *Gestalt Psychology, op. cit.*, p. 149.
[26] See *Information Storage and Neural Control, op. cit.*, p. 6.
[27] *Information Storage and Neural Control, op. cit.*, p. 5.
[28] *Ibid.*, p. 6.
[29] *Ibid.*, p. 6.
[30] Ludwig Wittgenstein, *Philosophical Investigations*, part I, paragraph 268. Wittgenstein is dealing with the concept of 'private ostensive definition'.
[31] *Information Storage and Neural Control, op. cit.*, from the article by Gregory Bateson, 'Patterns of human behavior', p. 174.
[32] *Information Storage and Neural Control, op. cit.*, p. 6.
[33] *Consciousness and Behavior, op. cit.*
[34] *Op. cit.*, p. 59.
[35] *Ibid.*, p. 80.
[36] *The Biology of Memory, op. cit.*, from the article by D. N. Spinelli, 'OCCAM: A computer model for a content-addressable memory in the central nervous system', p. 293.
[37] *Information Storage and Neural Control, op. cit.*, from an article by James G. Miller, 'The individual as an information processing system', p. 304.
[38] *Information Storage and Neural Control, op. cit.*, from the article, 'A logical calculus of the ideas immanent in nervous activity', p. 397.
[39] See, for example, in *The Biology of Memory, op. cit.*, the article by Bennet Murdock, Jr., 'Short and long-term memory for association', esp. p. 12, and the article by D. N. Spinelli, referred to above, pp. 298-301.
[40] *Information Storage and Neural Control, op. cit.*, article by Gregory Bateson, 'Patterns of human behavior', pp. 174 and 178.
[41] *Models of Human Memory, op. cit.*, the article of the same title, by Donald A. Norman, p. 2.
[42] *Ibid.*, p. 7.
[43] *Information Storage and Neural Control, op. cit.*, article by James G. Miller, 'The individual as an information processing system', p. 305.

[44] *The Biology of Memory, op. cit.*, article by Bennet B. Murdock, Jr., 'Short and long-term memory for associations', p. 12.
[45] *Ibid.*, his italics.
[46] *The Biology of Memory, op. cit.*, article by D. N. Spinelli, 'OCCAM', p. 295.
[47] *Brain and Behavior, op. cit.*, article by K. H. Pribram and W. E. Tubbs, 'Short-term memory, parsing, and the primate frontal cortex', p. 502.
[48] *Models of Human Memory, op. cit.*, article by Donald A. Norman and David E. Rumelhart, 'A system for perception and memory', pp. 19–20.
[49] *Ibid.*, p. 21.
[50] *Ibid.*, p. 21.
[51] *Ibid.*, p. 21.
[52] *Ibid.*, p. 22.
[53] *Ibid.*, p. 22.
[54] *Ibid.*, p. 23.
[55] *Ibid.*, p. 28.
[56] *Ibid.*, p. 35.
[57] *Information Storage and Neural Control, op. cit.*, article by James G. Miller, 'The individual as an information processing system', p. 304.
[58] *Ibid.*, p. 305.

PART III

TRACE THEORY AS PHILOSOPHY

PART III

TRACE THEORY AS PHILOSOPHY

We have seen that the only mechanistic theories which are at all plausible are trace theories. (The alternative would be a magical mechanism which produced, e.g., the 'Star Spangled Banner' with nothing at all to guide it each step of the way.) And, while trace theories strike most of us initially as being plausible, we have seen that trace theories, too, fail to explain memory. In Part I of this book, we saw that, even if trace theory could account for examples of *memory*, it leads to a problem when we consider examples of *imagination*. That is, we examined a typical case of musical memory. And we offered a trace-theoretical explanation of this musical example. The explanation ran into trouble when we considered cases where the piece of music was transformed from the original in various ways. What we found was that, even if trace theory provided a scientific explanation of the memory example, it failed to do so in the case where the piece of music was called to mind (imagined) in an altered form. In fact, what we saw was that trace theory committed us to the existence of a magical mechanism which was required to produce the music in its altered form. Thus, even if trace theory provides a scientific (causal, mechanistic) account of memory, it nevertheless commits us to an unscientific, magical explanation of the closely related powers of imagination. This commitment is, of course, intolerable. Trace theory is supposed to be science – not magic.

In Part II of the book, we developed a second, equally damaging criticism of trace theory. Considering the problem of trace retrieval, a process required in all trace theories, we saw that the task could be performed only by a homunculus. The alternative to this (unacceptable) possibility was the existence of an infinite regress of retrieval mechanisms. And we saw that these two possibilities (or impossibilities) are equivalent. To avoid confusion, we examined stimulus-response and information processing computer theories. This was done in order to show that neither of these two popular variants of trace theory is immune to the homunculus criticism.

The suggestion that trace theory is not really science, but philosophy, may strike the reader as sounding odd. Certainly the various forms of trace theory *seem* to be scientific. They deal with such subjects as neuron firings in the brain. And surely, the correct account of neuronal activity would be part of a scientific theory. Such a theory might be so technical and complex that it would be foolish and impertinent for a philosopher to attempt to evaluate its worth. Insofar as trace theory is supposed to be an account of the human nervous system, it is concerned with real structures in the human body. Thus, the success or failure of trace theory as an account of memory would seem to rest on empirical evidence, and not on logical or metaphysical subtleties. Yet, despite these considerations, I maintain that trace theory is seen in its clearest light when it is viewed as philosophy, not science. In the final part of this book, we shall see why trace theorists are engaged in philosophy rather than in science. Further, we shall see that the philosophical view which underlies trace theory is fundamentally misguided.

One preliminary consideration which may lead us to wonder whether trace theory is scientific or philosophical is the following. Many neuropsychologists might be surprised to find that their theories are not the recent inventions of scientists. The origins of trace theory predate the scientific and technological revolution by several thousand years. Plato, for example, sets forth a trace-theoretical account of memory in the *Theaetetus*. The following passage is taken from Cornford's translation.

Socrates: Imagine, then, for the sake of argument, that our minds contain a block of wax, which in this or that individual may be larger or smaller, and composed of wax that is comparatively pure or muddy, and harder in some, softer in others, and sometimes of just the right consistency.

Theaetetus: Very well.
Socrates: Let us call it the gift of the Muses' mother, Memory, and say that whenever we wish to remember something we see or hear or conceive in our own minds, we hold this wax under the perceptions or ideas, and imprint them on it as we might stamp the impression of a seal ring. Whatever is so imprinted we remember and know so long as the image remains; whatever is rubbed out or has not succeeded in leaving an impression we have forgotten and do not know.
Theaetetus: So be it.[1]

Whether or not this account is supposed to represent Plato's own ideas on the subject, there can be no doubt that what he refers to here is a typical trace-type explanation of memory. We can go further and say that the trace in this case would be of the static rather than dynamic type. (The reader will recall the distinction between these two variants, found in Part I, Chapter One.) As in all trace theories, it is essential that the trace (in this case the imprinted image) be a structural representation of the thing which is to be remembered. If, for example, I see Socrates and Plato coming down the street, then my block of wax must receive the imprint of two people. If my wax is not homogeneous in texture, then the image of Socrates might be stamped faintly on a hard spot, while the image of Plato comes through clearly and distinctly. Thus, when someone asks me who I met on the street, I might reply, "Oh, Plato and, er . . , someone else; I don't remember whom." In other words, the trace here must be isomorphic to what is remembered. Otherwise, how could I know (remember) whom it was that I met on the street? That, I take it, is the point of Plato's remark about the wax being pure or muddy, hard and soft, etc. It can be seen right away, too, that this attempt at explaining the power of memory fails to solve any problems. Possessing the image does no good at all unless I can recognize whom it is an image of. How can I know that it is an image of Plato? In order to do that, I must recognize it as an image of Plato. And in order to do *that*, I must already know what Plato looks like; I must already remember what Plato looks like. And how am I supposed to do that? This ancient version of trace theory leads us to the same dead end as its more modern counterparts. If I recognize the image by consulting some further image (or images) then how do I recognize those images? (The regress). And if, on the contrary, I can recognize the *image* without any further images, then why can't I recognize *Plato* in the same way, straight off and without any images at all? At some point I must recognize something *without* consulting some further guide. We can see, then, that the explanation set forth by Plato does not differ significantly from more modern trace theories. It is the same, right down to the criticisms which render it useless as an explanation of anything.

Plato's reference to a trace-type explanation of memory is certainly not unique in the history of philosophy. Thomas Aquinas, for instance, offers some considerations which show that this way of thinking had not gone completely out of fashion in medieval times.

> To receive and to retain are referable, in corporeal things, to different principles; for the moist receives well, but retains badly; and the contrary is true of the dry. Whence, since the power of sensibility is the act of a corporeal organ, it must be one power which receives sensory impressions, and another which retains them.[2]

Aquinas' suggestion that there are two distinct 'powers' – one which receives and one which retains, has an interesting parallel in a more modern theory which we have already referred to. In the article 'A System for Perception and Memory',[3] the authors propose a multi-stage memory process. The first stage consists in creating a sensory image from the incoming sensory input. At a later stage, another mechanism transforms this sensory image into a representation which is stored in the permanent memory.[4]

For a final example of trace theory put forward by a philosopher, we will look at some interesting passages in the beginning of Hobbes' *Leviathan*. In the first passage he defines sensation as a type of motion in our bodily organs, produced by other motions in external matter:

All which qualities called *sensible* (Hobbes' italics) are, in the objects that causeth them, but so many several motions of the matter, by which it presseth our organs diversely. Neither in us that are pressed, are they anything else but divers motions; (for motion produceth nothing but motion) . . . So that Sense in all cases is nothing else but original Fancy, caused, (as I have said) by the pressure, that is, by the motion of external things upon our eyes and ears, and the other organs thereunto ordained.[5]

Next, Hobbes defines imagination as decaying sense motions within us:

When a body is once in motion, it moveth (unless something else hinder it) eternally; and whatsoever hindreth it, cannot, in an instant, but in time, and by degrees quite extinguish it. And as we see in the water, though the wind cease, the waves give not over rolling for a long time after; so also it happeneth in that motion which is made in the internal parts of a man, then, when he sees, Dreams, etc. For after the object is removed, or the eye shut, we still retain an image of the thing seen, though more obscure than when we see it. And this is it, the Latines call *Imagination*, (Hobbes' italics) from the image made in seeing, and apply the same, though improperly, to all the other senses . . . *Imagination*, therefore, is nothing but *decaying sense*, (Hobbes' italics) and is found in men, and many other living creatures, as well sleeping as waking.[6]

Finally, Hobbes equates imagination with memory:

This *decaying sense*, when we would express the thing itself, we call *Imagination*, as I said before: But when we would express the *decay*, and signify that the sense is fading, and old and past, it is called *Memory*. So that *Imagination* and *Memory*, are but one thing, which for divers considerations hath divers names.[7]

So for Hobbes, memory consists of the decaying motions in our organs. These motions, though faint and 'obscure', are the same motions which, when they first occur, constitute sensation. Thus, we remember something only so long as we possess within us a pattern of motions which correspond to the thing we remember. What Hobbes is presenting here is nothing other than a primitive version of dynamic trace theory, a variant of trace theory in which the trace is not a static structure, but a dynamic pattern. More modern versions of the theory, which employ sophisticated terms like 'modulated carrier', are essentially no different from Hobbes' theory. The modern theories merely offer specific proposals as to the particular type of mechanism involved.

We can see, then, that trace theory is old, not new. It seems as if it has always struck people (at least some people) as being a very natural and reasonable way to explain memory. I offer these examples to show that trace theory is not an ultra-modern creation of western technological genius. Rather, it seems more correct to say that the trace theorists of today are carrying on in a philosophical tradition which goes back at least to Plato. It would, of course, be a mistake to assume that this fact (i.e., the fact that trace theory originated among philosophers) shows that trace theory is philosophy, not science. After all, science and philosophy were not necessarily thought of separately in times past. The rise of science (in the modern sense) is a relatively recent phenomenon. In Plato's time, philosophers were often concerned with questions which we now judge to be scientific, not philosophical problems. (There was, for example, speculation by philosophers about the shape of the earth.) So the mere fact that trace theory originated among philosophers does by no means show that it is philosophy rather than science.

But the issue certainly does not rest here. There is more to be said about the significance of our historical examples. In a sense, it would be misleading to call Plato's wax theory a *primitive* trace theory. It is only primitive in the sense that the particular mechanism proposed is crude. Plato did not, of course, know anything about neurons and computers, etc. So naturally, his trace theory does not contain any such references to sophisticated scientific concepts. But in essence, it does not matter whether the particular mechanism proposed is crude or sophisticated. (Plato was, presumably, speaking metaphorically anyway; he probably did not really suppose that peoples' heads contain actual blocks of wax with little images stamped upon them.) The point here is that modern trace theories differ from ancient philosophers' theories only with regard to 'hardware'. We can, for example, imagine Plato saying, "I have absolutely no idea as to exactly what sort of mechanism records the trace. Maybe someday someone will be able to build a model of one." And the modern trace theorists are engaged in just this task. They offer various suggestions for filling in the *details* of Plato's trace theory. Their trace theories are at bottom no different than Plato's. This is shown by the fact that these modern theories are susceptible to the very same criticisms which destroy the plausibility of Plato's speculations. It makes no difference whether the trace is supposed to be an image in wax, a 'modulated carrier', a computer program, or a magnetic field stored on tape. The criticisms offered so far, apply to all trace theories. Just to refresh the reader's memory, the criticisms are: In order for a trace mechanism to do its job, it must either (1) work by magic, (2) be operated by a homunculus, or (3) be one of an infinite series of mechanisms, each one operated by the one before it.

So the modern versions of trace theory in no way represent an advance over the earlier theories put forward by philosophers. They are the same. The common feature which exposes all of these theories to our criticisms is not based on the particular form the theory takes; the weakness is to be found in the very fact that they are all *trace* theories. The philosophical nature of trace theory is revealed when we realize that our general criticisms are *philosophical* in character. We are not maintaining that trace theory fails for *scientific* reasons. (We are not, for example, maintaining that trace theory fails because there are not enough neurons to be found in the brain, or something of the sort.) Rather, Parts I and II of the present work show that proponents of trace theory are committed to the existence of magical mechanisms, homunculi, and infinite regresses of retrieval mechanisms. And, of course, when we say that there *cannot* be an infinite number of retrieval mechanisms in Mr. A's brain, we are stating a philosophical truth, not a scientific one.

The preceding few pages may be summed up as follows. Trace

theory has its origins in ancient philosophical discourse. The trace theorists of today may be thought of as continuing in a tradition which goes back (at least) to Plato. The modern theories are essentially more detailed versions of their historical counterparts. Indeed, the very same criticisms which apply to the older versions are equally devastating when applied to the very latest, sophisticated versions. And, finally, these criticisms are philosophical, not scientific. These observations serve the purpose of filling out the claim that trace theory itself is, at bottom, philosophy rather than science. Nevertheless, even if these considerations are correct, I don't suppose that they settle the issue.

But then, they are only offered as a first step toward filling out the claim; there is more to be said on this matter. So far we have seen that mechanistic memory theorists are faced with a dilemma. For it seems, on the one hand, that a mechanistic theory of memory requires the existence of some form of trace – the (unacceptable) alternative is the existence of a magical mechanism. On the other hand, this requirement is the very feature which exposes all trace theories to the criticisms developed in the preceding parts of our examination. And, insofar as the criticisms are philosophical, trace theory fails for philosophical reasons. This conclusion is remarkable enough in itself, and probably comes as a surprise to those who see trace theory as a bona fide scientific attempt to explain how memory works. But it is not clear, at least to me, that this is enough evidence to support the claim that trace theory is essentially philosophy. It is true that the theory fails for philosophical reasons. But perhaps all this shows is that trace theory is philosophically confused science, mistaken science, but science nonetheless. And, if this were the last word on the subject, it might be a mere terminological dispute (whether a theory which fails for philosophical reasons can rightly be called a scientific theory).

But, in fact, it is not merely that the requirement of some trace exposes the theory to philosophical objection. In addition, I believe that the requirement itself is a philosophical one. That is, the idea of the trace is itself a philosophical concept. If this is correct, then we certainly have as good reason as we possibly could for maintaining that trace theory is essentially philosophy, not science. What we shall see in the final pages of this book is that the very notion of the trace springs from philosophical considerations. And further, the philosophy which forms the basis of trace theory is fundamentally misguided, confused, and even incoherent.

However, before we see just what is wrong with the notion of a memory trace, a very important problem must be dealt with. It might be supposed that my lines of attack on trace theory contain a serious flaw, a gap. It might be thought that the attack succeeds only against one particular type of trace theory, and that for all that has been said, there is a form of trace theory which is immune to the criticisms that have been brought forward. Imagine a critic taking exception to my procedure as follows: "The only type of trace theory which you have considered is an extremely rigid mechanistic, deterministic view. That is, the trace theorists you have attacked are ones who view human memory as a totally machine-like phenomenon. To such theorists, a person is nothing but a complex machine. However, this strict mechanistic view is not the only possible trace theory of memory. Couldn't there be a more moderate form of trace theory, one which was not so insistent in its demand for an absolutely machine-type causal explanation of memory abilities? Couldn't such a theory still allow for the existence of memory traces? It could be that memory traces do play *some* role in human memory, but that the rigid mechanistic theories you have examined are worthless. Thus, it may be that your attack fails, because you have failed to distinguish extreme positions from more moderate ones. Or, to put it differently, your attack has succeeded, but it is an attack only on *some* trace theories. Your criticisms don't work against a more moderate but still interesting claim that memory *includes* a trace storage mechanism, even though considerably more than that is needed to explain the existence of memory phenomena. You have, then, failed to show that the memory trace is a myth."

If this objection were justified, it would vastly limit the scope and power of my arguments against trace theory. It will therefore be necessary to reply to the objection in some detail. In the first place, it must be admitted that the target which this book has focused upon is indeed the sort of extremely mechanistic view mentioned by the objector. But even if this type of trace theory is extreme in one sense, that does not mean that such a view is uncommon. Rather, I think that this extreme form of trace theory is very commonly held. The view that memory is a machine-type phenomenon is often merely the tip of the iceberg. This view is closely linked with the view that people are nothing but machines; and this latter view is, in turn, linked to the view that science (biochemistry, etc.) holds the ultimate key to the mysteries of human behavior. Clearly, a major segment of our society does adhere to the kind of faith in science just alluded to. Scores, and perhaps hundreds of millions of dollars are being spent by universities, by the military, and by other governmental agencies, on the assumption that computers can be designed which will ultimately model the workings of the human mind. The foundation which obviously underlies such 'artificial intelligence' projects is the assumption that the workings of the human mind will turn out to be literally workings of some actual biochemical mechanism. So I don't think that the 'extreme' type of trace theory examined so far is an unimportant variation on an important theme. My own feeling is that this sort of view (the view that people are ultimately nothing but super-sophisticated biomechanisms) is the guiding light behind a great deal of thought, research and publication going on today.[8]

But my reply to the objection does not rest in any way upon the commonness or rarity of any particular sort of trace theory. For the arguments I have presented are applicable to all trace theories, and not just to some. I would attack any memory theory which included a trace storage mechanism of any sort. The views I am attacking are the ones which subscribe to the following principle: "For a person to remember *X* (whatever *X* may be), requires that the person possess a copy or trace of *X* (a trace which has the same structure as *X*)." Any theory which makes this claim is vulnerable to attacks like the ones brought forth in this book. In order to see this, let us consider the type of theory (a 'moderate' trace theory) which the objector has in mind, and compare it to the type of view examined so far. In Part II, Chapter Two, we looked at stimulus-response and information processing computer theories of memory. The point of that section was to show that, *no matter what* kind of machine is supposed to operate upon the traces, it just won't fill the bill. The complex creativity, imaginativeness, and variety of human memory phenomena (their *lack* of unity!), demand what a machine simply cannot produce. It matters not at all what kind of a machine we may dream up – or build.

In contrast to these 'hard' mechanistic theories, which we have already rejected, imagine a more 'moderate' trace theory: The brain contains a trace storage mechanism, a mechanism which is necessary for memory. But the theory recognizes that this by itself is not enough; rather, the trace is part of the data used by the person (by an intelligence), in order to formulate and execute memory-reactions such as the whistling of a tune. A theory of this type is still a trace theory – for a memory trace is required to enable the person to remember. But it is a *moderate* trace theory, for a trace storage and playback mechanism is not considered sufficient for memory. Instead, we need a person as part of the whole setup, a person who makes use of the data stored in the trace. Thus, the theory is only moderately mechanistic, since the trace mechanism only *partly* accounts for the appearance of memory phenomena. What are we to say about such a theory? Does it really avoid the criticisms levelled against against other forms of trace theory?

I think not. Bear in mind that the proposed theory claims to be an *explanation* of human powers of memory. Now take, as an example, Mr. A, who knows and loves most of Bach's music, including the '4b'. We shall assume that Mr. A's nervous system includes a trace of the '4b'. Suppose that Mr. A is placed in a situation in which he will have to make use of the trace. For example, Mr. B asks him, How does the '4b' begin?. And suppose that Mr. A promptly whistles the opening bars of the '4b'. According to the theory we are now considering, this display of memory ability was brought about in the following manner. (1) A trace of the '4b' was laid down in Mr. A's brain sometime in the past. (2) Mr. A, a person, an intelligence, calls upon his trace mechanism to give him the data necessary to perform the requested task. (3) The mechanism brings out the correct trace for inspection. (4) Mr. A is satisfied that the mechanism has indeed brought him the correct trace. (5) Mr. A proceeds to whistle the tune, using the memory trace of the '4b' as his guide. This process, or something roughly similar, is what our objector has in mind.

But a little thought suffices to show that such a theory provides no explanation of anything at all. We might, for example, ask the objector, How does the trace mechanism just happen to call forth just the right trace for inspection? (step 3). All the criticisms of Part II (the homunculus, the infinite regress of recall mechanisms) apply with full force here. On the other hand, our objector might say that we are mistaken here: "The trace mechanism simply *records* the trace. The mechanism does not *recall* the trace; rather, Mr. A does." However, such an answer will not do. How is Mr. A supposed to know which trace is the correct one? The only way he could know would be for Mr. A to *remember* what the trace of '4b' looks like. And here we see that we are caught in a regress again. In other words, a memory trace would only be effective when used by a person who is already possessed of a complete memory. So in this case we are still left with the question, How does Mr. A remember? (I pass over the obvious, but often unnoticed criticism that people don't generally engage in any process of searching through vast memory storage banks when asked to whistle tunes.)

Perhaps there is no need to elaborate upon this matter. It should be clear that, in this 'moderate' type of trace theory, the person, the intelligence, plays the same role as the 'executive programmer' in information processing computer theories. That is, Mr. A here functions as a little black box which does the remembering. And, if asked what is wrong with this notion, I remind the reader that the theory is supposed to offer an *explanation* of human memory. Yet what our 'moderate' theorist winds up saying is that the proposed memory process involves a person who operates the trace mechanism – a person with a memory! Once again we are dealing with a parody of a scientific explanation. Nevertheless, our objector might not be satisfied. He might say that the intelligence or person or soul which operates the trace mechanism is a very wonderful and powerful sort of being – a being which *can* recognize the right trace among many thousands laid down by the recording mechanism. But in that case, the entity in question must already have a full-fledged memory. So what is the trace mechanism supposed to be doing here? If Mr. A (or his intelligence) can recognize a *trace* without any help, then why can't he recognize a *tune* without any help? The trace here plays no part in the explanation; it is merely a smokescreen with a pseudoscientific odor. I end the digression by remarking that the 'moderate' trace theory offers no advantages over the more forthright mechanistic versions.

It is no accident that both theories wither under the same criticisms, for both sorts of theories share the same fatal weakness; they both require the existence of a memory trace. And, as I claimed just prior to the digression, it is the very concept of the memory trace which contains a fatal flaw. I said earlier that the trace is a philosophical concept, and that the flaw in this concept is a philosophical flaw. These claims are important, for they serve as justification for the central contention of this book – that trace theory is philosophy, not science, and that trace theory is worthless as an explanation of memory. With our digression safely behind us, we may now return to the questions: "Just what is wrong with the notion of a memory trace, and just why do I call the trace a *philosophical* concept?"

First, it should be pointed out that the necessity which seems to attach to the idea of the trace (discussed in Part I) is certainly not based on any empirical evidence. Rather, it seemed as if there *must* be a trace. It seemed as if we were forced to assume the existence of some trace. How else could someone (for example) hum correctly all the notes of the 'Star Spangled Banner'? The only answer that struck us as being at all plausible was that the person who calls to mind (hums, sings, writes down the melody from memory, etc.) the 'Star Spangled Banner' must possess a representation of it. How else could such a person get any of the details right? Whatever else may be said about this reasoning, it must be admitted that it is not based on any *evidence*.

All of the sophisticated experiments which have been performed by neurophysiologists and neuropsychologists have been evaluated by people who *assume* the existence of some sort of trace. Lashley, for example, found that he could remove something like 95% of the visual area of a rat's brain, and yet the rat would still react as it had been trained to do when presented with (e.g.) a green triangle. The removal of the 95% of the visual region didn't seem to affect the rat's memory for visual cues. This suggested to Lashley that there were no discreet locations for particular memory traces. That conclusion is, perhaps, justified. But then, Lashley went on to conclude that, since the memory traces didn't seem to be stored in any particular location, they must therefore be stored diffusely, at many places.[9] It would, of course, be consistent with the results of Lashley's experiment to conclude that, after all, we were mistaken in supposing that the rat's brain contained memory traces. But he never even considers this possibility. Now Lashley was not lacking in intelligence; quite the contrary. Yet he didn't even consider the possibility that there might not be any memory traces. I would suggest that he was in the grip of the powerful considerations which make it seem as if there *must* be memory traces. And this compulsion, whatever its nature, does not arise from any evidence. Lashley's experiment is a typical example of the research done by trace theorists. The experiments are designed only to test particular hypotheses about particular types of trace mechanisms. No evidence is offered for the existence of traces themselves. Rather, it is a matter of principle that there must be traces.

So the memory trace is not an object of scientific discovery. What, then, is it? Stripping away all particular suggestions about the nature of the memory trace (e.g., that the trace is dynamic, or static), we can characterize it in a very general way: The trace is something which is *isomorphic* to the thing remembered. We briefly examined the concept of ismorphism in Part I. But now is the time to look more closely at this notion. For it forms the heart of all trace theories. Indeed, all that a trace *is*, is something isomorphic to the remembered thing. It doesn't matter if the trace is dynamic or static, located in the brain or the spinal cord, discreetly or diffusely located, single or multiple; it must be *something* which is isomorphic to what is to be remembered. It is this concept of isomorphism which forms the essence of the trace. And isomorphism, as we shall see, is an incoherent philosophical concept. And because the trace is supposed to be *essentially* an isomorphic entity, the trace is *essentially* a confused philosophical notion. Thus, trace theory itself is shown to be philosophy, not science.

But what is wrong with the concept of isomorphism? When it is said that the trace is isomorphic to the thing remembered, what is meant is that, for each element of the remembered thing, there must be a corresponding element of the trace. If the remembered thing displays a pattern of arranged elements, then the trace must also contain elements which are arranged in a similar pattern. In other words, when we say that the trace is isomorphic to the remembered thing, we mean that the trace has a *structure* isomorphic to the structure of the thing to be remembered. The assumption here is that the thing to be remembered has a structure. Let me repeat that last sentence with a slightly different emphasis:

The assumption here is that the thing to be remembered has a structure.

Let us examine this notion of structure. We shall choose an example of something to be remembered, and ask what its structure is. Once we find out what its structure is, we shall thereby find out the structure of its corresponding memory trace. (We shall thereby find out the structure which the memory trace *must* have.) Let us choose as an example the 'Fourth Brandenburg Concerto'. We ask then, the absolutely crucial question, "What is the structure of the 'Fourth Brandenburg Concerto?' But does this question have an answer? Suppose we answer that the structure of the '4b' is composed of three movements, a serene allegro moderato, a poignant andante, and a brisk allegretto (or something of the sort)? This might be one reasonable answer. A music critic, for example, might describe it in this way. But suppose that someone (say an orchestral conductor) wanted to know what the structure of the '4b' was. The proper answer might be that the '4b' was composed of the following notes: (And then the conductor would be given a complete written score of the entire concerto.) That might be another answer to the question, "What is the structure of the '4b'?" But suppose that a physicist asked the question, wanting to know the acoustic structure of the '4b' in terms of sound frequencies at differing pressures, plotted as a function of elapsed time. In fact, we can imagine that a physics professor asks this question in a class. If a student answered that the '4b' was composed of three movements, a serene . . . (etc.) the physicist would say, "No, no! I mean, what is the acoustic structure, in terms of sound frequencies, pressure, and so forth." In other words, the music critic's answer is the answer to a completely different question than the physicist's. We can imagine a historian of music answering the question this way: "The '4b' is composed of Bach's genius, the musical tradition of baroque composition, and the patient skill of the thousands of musicians and conductors who have performed this work for centuries." Or again, we might say that the '4b' is composed of a first recorder part, a second recorder part, a first violin part, a cello part, etc. No one answer has any more claim to being the 'right' answer than any other. They are all possible answers, depending upon who is asking them, and why. Another way of putting this point is to say that there are actually a number of different questions which might be asked by saying, "What is the structure of the '4b'?". Thus, when the physicist asks the question to his class, a clever student might reply, "Do you mean to ask, "What is the *acoustic* structure of the '4b', professor?". Presumably there would be no need to say this, since most students would realize that a physics class is no place to talk about poignant andantes.

But what is the point? The point is simply that, in order for the question, "What is the structure of the '4b'?" to make sense, it must be asked by someone who has in mind a particular sort of structure. The preceding sentence might be misconstrued, so let me try to make the point another way. We said that there are a number of different meanings which our question might have. (Perhaps an indefinite number.) But in order for it to be meaningful, it must have *some* particular meaning. That is, the question must be asked by someone, and it must be asked *to* someone. The question must have *some* context which makes one sort of answer reasonable. Standing all by itself, in a vacuum, as it were, the question wouldn't have any one right answer. That is, in a vacuum, the word 'structure' means nothing. In order to see this, imagine the following scene: The trace theorist asks, "What is the structure of the '4b'?" We reply, "Do you mean the acoustic structure, the arrangement of notes, the arrangement of movements, or what? What do you mean by 'structure'?" The trace theorist replies, "I just mean the structure. I mean *the* structure. What is *the* structure of the '4b'?" The word 'structure' here is meaningless. It might mean anything – or nothing. Take another example. What is the structure of an ashtray? In chemistry class we could say it is leaded glass. In geometry class we could say it is cylindrical, in art class we could say that its structure is typically American motel, circa 1970. But what is *the* structure of the ashtray? The reason why the question is crucial is that its answer will also tell us the structure of the memory trace of the ashtray (which is isomorphic to the structure of the ashtray). But if the idea of structure here is empty nonsense, then so is the idea of the trace which *essentially* has that structure.

These remarks about structure also apply to the idea of an *element*. When we say that two things have the same structure, we mean, for example, that the elements of the two things are arranged in the same way. But what elements are we talking about? What are the elements of the '4b'? Physical vibrations in air? Time? Movements? Violins? People? Bach? Silence? History? What are the elements of an ashtray? Glass? Silicon and oxygen? Circles and straight lines? Matter and form? Colors? Mass? Beauty? Sharp points? Nicks and scratches? We haven't a clue as to what is meant here. At the heart of trace theory, then, we find the idea of isomorphic structure. But this concept is used by the trace theorist as if it made sense all by itself. (As if it made sense for me to ask about the structure of an ashtray, without my having any idea whether I mean chemical structure, geometrical structure, artistic structure, or anything else. And if *I* haven't any idea what I mean by 'structure', or 'element', then who does?)[10]

This confused and incoherent notion of structure, which lies at the heart of all trace theory, brings with it other confused notions. Wolfgang Köhler, for example, is interested in the problem of association between memory traces. (We talked about association in Part II.) We will look again at a passage presented earlier. But this time, we will also examine the passages which follow the one we saw earlier. Here is the passage we examined before:

> What does recognition mean? It means that a present fact, usually a perceptual one, makes contact with a corresponding fact in memory, a trace, a contact which gives the present perception the character of being known or familiar. But memory contains a tremendous number of traces, all of them

representations of earlier experiences which must have been established by the processes accompanying such earlier experiences. Now, why does the present perceptual experience make contact with the trace of the *right* (Köhler's italics) earlier experience? This is an astonishing achievement.

Köhler goes on to explain this achievement:

Nobody seems to doubt that the *selection* (Köhler's italics) is brought about by the similarity of the present experience and the experience of the corresponding earlier fact. But since this earlier experience is not present at the time, we have to assume that the trace of the earlier experience resembles the present experience, and that it is the similarity of our present experience (or the corresponding cortical process) and that trace which makes the selection possible.[11]

For Köhler, then, the right trace is chosen on the basis of its similarity of structure with the memory which it represents. Thus, for example, when we hear a piece of music and recognize it as the '4b', we recognize it on the basis of the similarity of structure between the present rendition of the '4b' and the memory trace of an earlier rendition. The point we made in Part II was that such recognition would require a homunculus. But the point we make here is that Köhler is assuming that the '4b' has *a* structure which is similar to *the* structure of the memory trace. And we must ask, "What do you mean by 'structure', and in what sense are you using 'similar'?"

He goes on to a further discussion of the importance of similarity:

When subjects learn a series of items, the effect of the learning is disturbed if, between the learning and the test of recall, the subjects have to deal with items of a similar kind.[12]

Köhler is not alone in this observation. W. R. Russell makes a similar point. He is concerned with memory traces which represent skilled bodily movements. The general claim he makes is that learning a particular skilled movement can interfere with one's ability to perform other, 'closely related' skilled movements. He offers the example of a golf swing, saying that one can learn a new way of swinging the club, and that this new learning experience might make it difficult to perform the older style of golf swing. This learning experience would not, for example, make it more difficult to play a particular tune one had learned on the piano. I think his observation of this phenomenon is quite correct. But his explanation of it is misguided. He suggests that the neural circuits which store the trace of the golf swing are disturbed by other circuits which store the traces of similar skilled movements. But in what sense are two styles of golf swing similar? There might be *many* senses in which they are similar. Both involve, for example, grasping something with the hands and swinging it toward the ground. But, of course, there are many senses in which two golf swings might be dissimilar. For example, one might be flamboyant, and the other quite precise and understated. In this sense, we would be correct in saying that a flamboyant golf swing is similar to a flamboyant and romantic rendition of some piano tune. And we would be correct in saying that this flamboyant golf swing was *not* similar to a very precise and reserved golf swing. In fact, we could even imagine someone whose golf swing affected his piano playing. Imagine that someone with a very reserved golf swing learns to play some flamboyant romantic rendition of a piano piece. Suppose that he decides that the romantic style really embodies an admirable attitude toward life in general. The next time he goes out to play golf, he steps up to the first tee, saying to himself, "Beethoven is where it's at." He raises the club in an unheard-of high arc, and amazes his friends by smashing a tremendous drive right down the middle of the fairway. Question: Is a golf swing more similar to another golf swing than to a piano rendition? Reply: What does 'similar' mean here? The same reply is appropriate when we are asked about the *structure* of a golf swing. – What does 'structure' mean here? And, as Wittgenstein says in paragraph 47 of the *Philosophical Investigations*, our reply is really a rejection of the question.

Another incoherent concept used by trace theorists to describe traces is the idea of 'being related', or 'being relevant'. This idea is closely linked with the idea of 'similarity' which we just examined. For example, consider the statement that "Memory of a particular thing can be interfered with by the learning of some other, similar thing." Another way of putting this is to say that "Memory of a thing can be disturbed by the learning of some new, closely related thing." This idea is just as empty and non-sensical as the idea of similarity, as it was used on the preceding few pages. It makes no sense to ask whether two things are related, unless we have *some idea* as to what sort of relation we mean. Yet we find this nonsense over and over again in trace theories. Lashley, for example, thinks that traces are organized into groups. These groups are formed according to relations of relevance or 'relatedness'. He offers an example of two types of information which are assumed to be unrelated, and are therefore presumed to be stored in two different groups of memory traces. His example: chess strategy is unrelated to knowledge about human anatomy.[13] A final example of this empty, contextless use of the idea of being related is found in an article to which we had occasion to refer earlier. The authors suggest that an item in the long term memory store might be represented "by a list of its relationships with other material."[14]

Our results can be summarized in the following way. Trace theorists maintain that someone recognizes (e.g.) the '4b' by recognizing that the structure of the present musical experience is the same as the structure of the trace of the '4b'. We pointed out in Part II that this move (from recognizing x to recognizing the structure of x) does *not* contribute anything toward an explanation of recognition, memory, learning, or anything else. For recognition requires a homunculus with a complete memory. This is true whether the homunculus is supposed to recognize the '4b' or whether the homunculus is supposed to recognize the structure of the '4b'. But now we have seen, too, that the trace theorist's idea of 'structure' is confused and incoherent. Since the essence of the trace is to be something which possesses a structure in just this sense, our criticism strikes at the heart of trace theory, exposing it as being essentially confused.

We would, I think, be justified in dismissing this notion of *the* structure, now that we have seen that first, it is utterly confused and even incoherent, and second, that it lies at the heart of trace theory. The task we set ourselves at the outset of this study has been virtually completed. But, by pursuing this idea of structure a bit further, I think we can perhaps gain some valuable insight into the origins of this (misguided) way of thinking. The trace theorist's notion of structure springs from a philosophical source which covers a much wider area than just theories of memory. So, in order to follow it to its source, I propose that we consider the trace theorist's notion of structure as if it were legitimate. We have, after all, used this tactic in our earlier examination of the theory.

Suppose, then, that it made sense to speak of *the* structure of (e.g.) an ashtray, or the '4b', without regard to any particular kind of structure. What sort of thing would this structure be? That is, what sort of things could we say about it? Let's take the '4b' as an example. What, then, can be said about *the* structure of the '4b'? Well, we know that it is this structure that we recognize when we recognize something as being a rendition of the '4b'. When we say, "That music is Bach's 'Fourth Brandenburg Concerto'," (and when we are correct) then we are right because the music in question is music which has the structure of the '4b'. So, for example, any rendition of the '4b' must possess, or display this structure.

But this structure is the same structure which is to be found in our memory trace. Indeed, it is just this structure which makes our trace a trace of the '4b', and not, for example, a trace of the 'Star Spangled Banner'. Further, when someone owns a written score of the '4b' that person owns something which has the same structure as the '4b'. That is what makes it a score of the '4b', and not of some other piece of music. And what makes a phonograph record a record of the '4b', is the fact that the grooves and bumps on the record embody the structure of the '4b'. Our conclusion from all this is that the structure, as required by the trace theorist, is not composed of acoustic vibrations, brain cells, paper and ink, or record vinyl. Rather, all of these things (the record of the '4b', the musical performance of the '4b', the written score of the '4b', and the memory trace of the '4b') possess the same structure – the structure of the '4b'. All these things have this structure in common. The thing that they have in common is *the* structure of the '4b'. And we must remark that this structure is much more abstract than any of its particular embodiments.

In fact, the structure of the '4b' will turn out to be very abstract indeed. At this point we might suppose that, although the structure of the '4b' may be embodied in different materials (on records, in brains, etc.) nevertheless it is always possessed of certain characteristics which do not vary from one particular embodiment to another. For example, consider the '4b' as an acoustic performance, and as a phonograph record. While one consists of bumps in vinyl, the other consists of vibrations in the air. But we might suppose that both of these embodiments of the structure of the '4b' are characterized by certain constant characteristics. For example, there will be a certain distance between bumps on the vinyl surface. And this *distance* between bumps will correspond to the *time* which elapses between the vibrations of the acoustical performance. Although one embodiment consists of bumps in vinyl, and the other one consists of vibrations in the air, nevertheless there will be an isomorphic correspondence between the two embodiments. Thus, although the two embodiments are in different material forms, nevertheless there *will* be a characteristic pattern. And that pattern is shared in common between the bumps on the record and the acoustical vibrations. And we might suppose that it is this characteristic pattern which identifies both embodiments as being embodiments of the structure of the '4b'. Thus, the structure is abstract, in that it can be embodied in vinyl bumps or in acoustic vibrations. But at least there is something which remains constant. And now we have, so to speak, purified the structure of the '4b'. We are not bothered by the fact that it is abstract. After all, *the* structure of the '4b' is just what is left over when we get rid of the vinyl, the brain cells, and all the other accidental accompaniments. The common element in all these isomorphic embodiments is *the* structure of the '4b' itself. Of course it is abstract. Who would think that *the* structure of the '4b' would be made of some particular material, anyway?

But just what is left after this process of abstraction? Forget about the vinyl, the vibrations, the brain cells, etc. Will there really be an abstract, characteristic pattern to all of these things? What would an example of a portion of this pattern be like? Well, the sort of abstract pattern we referred to in the preceding paragraph would be something like the following: In the phonographic embodiment of the '4b', we might find that the first three notes of the '4b' are represented by three vinyl bumps, spaced equidistant. These three bumps would correspond to three sets of vibrations in the acoustic embodiment of the '4b'. Now, the fact that the three bumps are equidistant would correspond to the fact that the three sets of vibrations are equidistant in time. Abstracting from this, we see that there is a common pattern between the two embodiments of the '4b'. And whether we render this pattern in terms of time between vibrations, or in terms of distance between vinyl bumps, the pattern remains the same. And it is this abstract, mathematical or logical pattern which is *the* structure of the '4b'. In our example of the first three notes, we could say that the 'distance' (whether in time or space) between the first element and the second element would be equal to the 'distance' between the second element and the third. And this would be true whether we are referring to the actual distance between two vinyl bumps, or whether we are referring to the 'distance' in time elapsed between two acoustic vibrations. This equality has different representations in different embodiments of the '4b'. But the fact that it *is* an equality does not change. That would be part of *the* structure of the '4b', however embodied. This sort of abstract pattern is an example of what would be left over when we tried to abstract the common element from different embodiments of the '4b'. And it is this abstract pattern which makes something into an embodiment of the '4b'.

But, if we consider different examples, we will see that actually there is no such constant pattern to the different embodiments of the '4b'. Consider different orchestral renditions of the '4b' during which, for example, the first note of the first phrase is held a bit longer than the second note. That is, there are renditions of the '4b' in which the time elapsed between the first note and the second is *not* equal to the time elapsed between the second note and the third. And, though someone might wish to say that such a rendition of the '4b' would be in poor taste, nevertheless it must be admitted that it would be a rendition of the '4b'. And we would recognize it as such. Thus, according to the trace theorist, when we recognize this romantic rendition of the '4b' as being a rendition of the '4b', we must be recognizing in it *the* structure of the '4b'. And so, in this case, *the* structure of the '4b' is embodied by something that does *not* display the particular pattern of equality which we mentioned before. So even this particular abstract mathematical pattern is not part of *the* structure of the '4b'. Rather, *the* structure of the '4b' must be something even more abstract. It must be a structure common to all the different renditions of the '4b'.

The effect of the preceding paragraph is to show that a certain type of pattern, one which at first seemed abstract enough to qualify as part of *the* structure of the '4b', is, in fact, not suited to the purpose. Rather, that particular abstract mathematical pattern is found only in *some* renditions of the '4b'. But, of course, this result can be applied to more of the '4b' than just the first three notes. There will be all sorts of differences in the timing between the notes of different renditions of the '4b'. Some will be faster, some slower. Some will have a quick first movement, a slow second movement, and so on. There are an incredible number of different possible interpretations for each *phrase* of the '4b'. Yet all of these renditions must display the abstract structure which is *the* structure of the '4b'. At this point, whatever is to qualify as being *the* structure will have to be very abstract indeed. Consider: *The* structure of the '4b' will be something possessed in common between the following two embodiments. The ones I have in mind are first, a phonograph recording of a very romantic interpretation of the '4b', and, second, an acoustic performance of a quick-tempoed and precise rendition, perhaps done on a Moog synthesizer. In the precise acoustic performance, we find sets of vibrations in air, many of them separated by equal time intervals. In the romantic phonograph record, we find sets of vinyl bumps, many of them separated by distances which are not equal. In fact, the unequal distances between the bumps on the record are to be compared with the equal time intervals in the acoustic performance. Whatever they have in common will, perhaps, be part of *the* structure of the '4b'. But what do they have in common? Not much.

The reader will by now recognize that we are engaged in a process of 'whittling down' the notion of *the* structure of the '4b'. This process is quite similar to the process of 'whittling down' the trace, which was done in Part I, chapter 2. In that chapter, we found that each feature which we examined *need* not have been part of the trace of the '4b'. But here, we see that each feature which we examine *cannot* be part of *the* structure of the '4b'. Take, for example, the abstract mathematical feature of 'equality of separation between the first two pairs of elements of the '4b'. (That might be one clumsy way of describing the abstract feature in common between the phonograph record and the acoustic performance.) We found that even this level of abstraction will not do. For there are (e.g.) phonograph records which do not display this pattern, and acoustic performances which do. *The* structure of the '4b' is what they have in common. It enables us to recognize both as being renditions of the '4b'. So the abstract pattern (which belongs to one rendition and not to the other) obviously is not part of *the* structure of the '4b'.

What features are left, which the different renditions of the '4b' might possibly have in common? Different renditions of the '4b' (different embodiments of *the* structure of the '4b') can be played at differing loudnesses, at differing speeds, in differing rhythms (rhumba rhythm, for example), etc. Yet they will all be embodiments of *the* structure of the '4b'. There is no feature which must be common to all the things which we rightly call renditions of the '4b'. Let us hark back to an example presented in Part I. Imagine someone humming an imitation of a trumpet, playing the '4b' too fast, in a rhumba rhythm, extremely loud, with a couple of trills thrown in for embellishment, and perhaps the first note of each phrase an octave higher than it usually is. (I myself can actually do this in mambo rhythm; I don't know the rhumba.) Such a rendition of the '4b', while it may be unorthodox, would definitely be recognizable to many people who have had any musical training, or who are naturally musical. One of those people could correctly say, "Why, that outlandish fellow is doing a nightmare version of the 'Fourth Brandenburg Concerto'! I would know that piece anywhere." And the trace theorist would explain the ability to recognize the '4b' in the following way: "When someone recognizes the '4b', they recognize *the* structure of the '4b'. That is, they recognize that the present musical performance has the same structure as the trace of the '4b'." And by way of reply, we ask the following question: "What is *the* structure of the '4b'?" The trace theorist replies, "It is what is found to be in common between different renditions of the '4b'." And, of course, this structure must be found not only in *acoustic* renditions of the '4b', but also in all things which embody *the* structure of the '4b', including phonograph records, memory traces, etc. And so now, we ask the trace theorist, "What *do* different renditions (embodiments) of the '4b' have in common?" For example, what is there in common between the following three things, all of which are embodiments of *the* structure of the '4b'? (1) Our outlandish trumpet imitation; (2) A phonograph record of a romantic rendition of the '4b'; and (3) A memory trace of the written score of the '4b'. What on earth do these three things have in common? What pattern, however abstract, could be found in all three of these embodiments? I can't think of anything which might serve as a common pattern between these three embodiments of the '4b'. But if someone can, it would be an easy matter to imagine some other embodiment which did not display this pattern.

At this point it must be admitted that there are some philosophers who would balk at the preceding few pages. They would probably be unhappy with my method of showing that there is no common structure to be found among the various renditions and embodiments of 'The Fourth Brandenburg Concerto'. Since the point is an important one, it would, perhaps, be wise to bring out the objection, and to deal with it now. The objection could be couched in the following terms. "You are playing fast and loose with the criteria of identity for Bach's '4b'. Though one may play the '4b' in all sorts of ways, there are limits to the changes one can make in the *notes*. If these limits are exceeded, one is no longer playing the '4b'. For example, your trumpet mambo version is not really the '4b', but a *parody* of the '4b'. Thus, your set of examples is illegitimately large. If we confine our search to *bona fide* versions of the '4b', we will surely find a common structure among them. Indeed, we already know what this common structure is. It is nothing else than the *tune*, the dominant melodic line of the '4b'. And what is the tune? Well, it is a succession of notes related to one another, a succession of notes which is heard as a single pattern."

The reply to this objection must be made in two parts. First, it seems to me that the notion of a parody is incorrect here. It is not the structure of a particular rendition of the '4b' which makes that rendition into a parody of the '4b'. Consider a rendition of the '4b' done by an orchestra of kazoos and tubas. That would be a parody. Nevertheless, the members of the orchestra might play their parts paying careful attention to the score and to the conductor. This scrupulousness and attention to detail would not prevent the performance from being a parody of the '4b'. Indeed, it might be that the more painstaking and meticulous the preparation and rehearsals, the better parody it would be. Further, although such a performance would be a parody to us, I can imagine a culture which traditionally performed such baroque music on kazoo and tuba. After all, most orchestral renditions of baroque music are not done with the original instrumentation. For example, piano is often substituted for harpsichord. And while this doesn't strike *us* as odd and ridiculous, it might very well have seemed so to listeners in (say) the early 19th century. There might well be people who regard a precise rendition of the '4b' done on a Moog synthesizer as an obscence parody of the '4b', worse, in a way, than the mambo trumpet version. The idea of a *parody*, then, is not the idea of something with certain structural distortions. (Although, of course, *some* parodies do qualify as parodies because of deliberate distortions of e.g., acoustic structure.)

So much for the idea of a parody. But with this portion of the objection dealt with, it is time to answer the more serious part. First, I must admit that the trumpet mambo version of the '4b' *is* a parody. But that fact does not warrant our excluding this admittedly extreme example from consideration. Why not? Because the trace theorist purports to explain how it is that we recognize (among other things) that a certain orchestral performance is a parody of 'The Fourth Brandenburg Concerto' and not (e.g.) a parody of 'The Fifth Brandenburg', or, for that matter, of 'The Star Spangled Banner'. Let us recall the manner in which the trace theorist would explain this. My ability to say (correctly) "What an awful thing to do to the '4b'!" is to be explained by the fact that my brain contains a trace of the '4b'. And the only reason why I recognize that it is the '4b' which is being parodied, is that there is a structural isomorphism between my memory trace and the present acoustical performance. The fact that the present performance is a parody of the '4b' is irrelevant. Thus the trace theorist *is* committed to the search for the abstract common structure to be found in *all* embodiments of the '4b'. This will include even the mambo trumpet version. And, as we have already seen, there *is* nothing in common to be found among all these embodiments. We will not, for example, find any "succession of notes related to one another . . .", as suggested by our hypothetical objector.

With this objection out of the way, let us now resume our examination of the notion of *the* structure of the '4b', *the* structure which, so to speak, makes the '4b' what it is. The idea of *the* structure of the '4b' is simply the idea of some abstract pattern which is common to all embodiments of the '4b'. But, as we see, there *is* no common pattern to all of the myriad things which embody the '4b'. Yet it is the assumption of trace theorists (and others) that there *must be* something which is *the* structure of the '4b'. Jerry Fodor, for example, realizes that there is a problem in determining just what abstract pattern is common to all acoustical performances of a particular tune. But he just goes on to assume that there must be *some* abstract pattern common to all of them. The following passages are from his book, *Psychological Explanation.*[15]

> . . . consider, that is, what a person has to know in order to be able to recognize renditions of 'Lillibullero'.
> It is clear, in the first place, that the set of events that one is capable of easily recornizing as a performance of 'Lillibullero' need not have any distinguishing acoustic characteristics. In fact, no two members of that set need have such characteristics in common. For one can recognize the tune when it is played on a warped record, transposed, played as a waltz, played as a march, and so on and on. (pp. 24–5)
> The present example is, of course, in no way special. A large number of the things one can recognize, such as shapes, tunes, sentences, and faces are drawn from stimulus domains that have quite complex mathematical structures. (p. 25).

Like the requirement for the existence of the memory trace, the requirement for the existence of *the* structure is not based on any evidence. Though Fodor cannot suggest any plausible candidate for *the* structure, he doesn't hesitate to assume its existence.

> Any serious attempt to construct a viable psychological theory of perception . . . would have to account for the fact that training often generalizes to objects that may be only quite abstractly related to the trained object. I cannot imagine how this is to be done unless it is assumed that the concept you have of a face, or a tune, or a shape . . . includes a representation of the formal structure of each of these domains and that the act of recognition involves the application of such information to the integration of current sensory inputs.[16]

Fodor assumes (as do all trace theorists) that recognition (e.g.) of the '4b' requires that the recognizer possess a representation of *the* structure of the '4b'. But we have tried to show that this requirement is not based on evidence. Rather, the evidence shows that there *is* no such thing as *the* structure of the '4b'. The only sort of thing which might qualify as *the* structure of the '4b' would be, for example, the arrangement of the ultimate elements of the '4b'. But we pointed out earlier that it makes absolutely no sense to ask "What are the elements of the '4b'?" when the question is asked in a vacuum. Nor does it make sense to ask, "And also, what is the structure which these elements are arranged in?" Yet the trace theorist assumes that these questions do make sense. Indeed, the trace is essentially nothing but an embodiment of 'structure' in just this sense (which is no sense at all).

But why does there seem to be a natural urge to assume the existence of something called *the* structure of the '4b'? For example, why does Fodor, by his own admission, find himself unable to imagine that things might be different? It is his assumption that things (e.g., the '4b') have a structure, which leads him down the blind alley of trace theory. It is the assumption that things have a structure, which leads him to assume that the only way people can recognize a thing is to recognize *the* structure of that thing. And we have criticized this concept of *the* structure of a thing. But what sort of philosophical view is this? One way to characterize the view would be like this: When someone says that a particular piece of music is the '4b' (and when that someone is *right*), then there must be something about the music which makes it the '4b' (or which makes it correct to say that it is the '4b'). And, whatever that something is, we shall call it *the* structure of the '4b'. A further assumption is this: When someone says that two particular pieces of music are both renditions of the '4b' (and when that someone is *right*), then there must be something about each of those pieces of music which makes them both renditions of the '4b'. Indeed, the assumption is more specific than that. The assumption is that whatever makes it right to call both pieces of music renditions of the '4b', must be the same in both cases. Another way of putting this is to say that there must be some *one thing* common to all renditions of the '4b', which makes them all renditions of the '4b'.

And, of course, this requirement is not meant to apply only to the '4b', or to pieces of music. It applies to ashtrays, chairs, songs, people, nations, life; in fact it applies to the universe and everything in it. That is, there must be *something* about ashtrays that makes them all ashtrays. (And there must be something about all games that makes them games.) And this something will be called *the* structure of ashtrays, *the* set of things common to all ashtrays. Everything in the universe has *a* structure. In fact, we might say that the universe itself has *a* structure (perhaps the structure formed by the set of all particular structures of particular things).

This then is the philosophical view which forms the very heart of trace theory. And our criticism was twofold. First, we said that this idea of structure is an incoherent one. The trace theorist has no idea what might be meant by *the* structure of a thing (or *the* ultimate elements which make up that structure). Our second criticism sprang from our examination of the '4b'. We found that there *is* nothing common to all the various renditions of the '4b'. And this comment applies to ashtrays, too. There *is* nothing which all ashtrays have in common, which makes them all ashtrays. But the idea that there *is* something in common is nothing but the old philosophical idea of *essence*. Another way to put it is to say that the thing which is common to all ashtrays, which makes them ashtrays, is nothing but the abstract Platonic Form of ashtray. Yet another guise the idea appears in is that of the 'token-type' distinction. All particular ashtrays are tokens of the same type. That is what they have in common. And, of course, when we correctly say that two things are tokens of the same type, there must be something common to all of the tokens – something which makes them all tokens of the same type. Though the trace theorist cannot *find* anything common to all the things we (*correctly*) recognize as ashtrays, nevertheless there must be something, however abstract, which enables us to recognize them. This abstract 'something' is what the trace theorist calls *the* structure of ashtrays (*the* structure of the '4b', *the* structure of games, etc.). The fact that each ashtray possesses this form makes it correct to call some things ashtrays, and incorrect to call other things by that name. And our possessing a *trace* which displays this structure is what makes it possible for us to *recognize* (remember, know) that something is an ashtray. Or so the trace theorist believes. But we have seen that the trace theorist is mistaken. First, there just *is* nothing in common between all ashtrays, or between all the different renditions of the '4b'. Second, the idea that there *is* something in common is the idea of *the* structure. And this concept (which is what Wittgenstein calls a 'super concept') is incoherent, as we have seen. Since the trace is actually nothing *but* an embodiment of the essence, or Platonic Form of a thing, we are forced to conclude that trace theory is, at bottom, philosophy, not science. This concludes our examination of trace theory.

There are a few points which it would be wise to mention, in order to avoid some possible confusions. First, when we say that trace theory is nothing but philosophy, we do not ignore the fact that empirical experiments have been performed in the name of trace theory, by trace theorists. Indeed, some of these experiments may prove valuable in terms of the sciences of neurophysiology, neurobiology and neurochemistry. For example, it is not inconceivable that some of the trace theory-inspired experiments on brain function may someday help us to cure amnesia. But by now the reader is in a position to see that the cure for amnesia will not be a matter of restoring lost memory traces. That is nonsense. The purpose of this entire work has been to show precisely what kind of nonsense it is.

Next, the reader might be puzzled by a point made earlier. We said that there is no common feature which makes (e.g.) ashtrays what they are. What, then, makes it right for us to call some things ashtrays, and wrong to call other things ashtrays? Is there no right and wrong here? The answer is that there is no *essence*, Platonic Form, or 'structure' which is common to all the things we rightly call ashtrays. Rather, there are a great number of different features which make it right for us to call something an ashtray. Ashtray #1 might be an ashtray because it is made of some material which won't catch fire or melt, and is of a convenient size and shape. Ashtray #2 might be made of waxed paper, but it is of a convenient size and shape. It will serve quite well with a bit of water in the bottom. Ashtray #3 might also be of convenient shape, but it may be eight feet long – a giant ashtray in an art store. (Price: $5000). Another ashtray (#4) might be too small, with almost no concavity to hold ashes; but we found it on the night table in our hotel room. The name, 'Hotel Pierre' is painted on the bottom. We know it is an ashtray. Ashtray #5 is on a new streamlined train. It is in the shape of a funnel, with a cover. When we press a button, the cover slides back, revealing the railroad tracks below. Ashtray #6 is made of inflammable plastic. But it is intended to be used as a movie prop, in a James Bond movie. And so on. Some ashtrays are called ashtrays merely because they are used for the purpose of holding ashes. But how about an ashtray in a painting? In some cases we can point to a feature of the ashtray itself; sometimes we point to some feature of the situation in which it is found. We might even point to a custom or social convention. An example of this might be as follows. We walk into a store selling Turkish souvenirs. On the floor is what seems to be a spittoon. We say to the proprietor, "How much is that spittoon?" He answers, "No, no. That is not a spittoon. It is an ashtray. In Turkey we consider it a capital offence when someone spits. You will never see a spittoon in our store'" What makes something an ashtray will depend on the circumstances, which make it correct to call something an ashtray. Ashtrays form what Wittgenstein calls a *family*. Why should it be thought that there is any *one* thing which makes all ashtrays what they are?[17] [18]

Where, then, do we stand with regard to the problem of memory? In the first place, it is now clear that trace theory does not offer a scientific explanation of the phenomena of human memory. Indeed, trace theory offers no explanation of anything at all. It is not even a first step toward an understanding of human memory. Earlier on, we saw that trace theory was the only form which a mechanistic, causal account of memory could plausibly take. As the reader may recall, our considerations were simple. Suppose Mr. A calls to mind a detailed visual image of the Mona Lisa. Now the trace theorist assumes that this visual image is created by a step-by-step, determined causal process. Each feature of the memory trace causes some aspect of brain function to change, thus resulting in the formation of a memory image. Any mechanistic, causal account of memory which denied the existence of the trace would face a grave objection. Such a theory would require the existence of a mechanism which, with nothing to guide it, somehow managed to put in the right details. At each step, this machine would put in the right colors, the right lines, etc. But, of course, no machine could do this. Or, to put it another way, any machine which *could* do this would have to operate by magic. And a theory which required the existence of a magical mechanism would be a parody of a scientific, mechanistic, or causal theory. So trace theory is the only possible mechanistic theory of memory.

But our examination raised objections against trace theory. In fact, in Part I, chapter 1, we saw that a memory trace theory would require the existence of the very sort of magical machine described previously. This machine would be needed to explain the fairly common capacity to call to mind drastically altered images of things we have seen and heard. Thus, trace theory is faced with the very same grave objection which destroys the plausibility of a non-trace mechanistic theory. (That is, trace theory, too, requires the existence of a magical mechanism.)

After having examined and rejected the idea of the memory trace, we must now return to two objections raised in Part I.* Our imaginary critic protested that there are a number of independent reasons for insisting that traces *are* necessary for memory. First, if there are no memory traces, this would obliterate the distinction between memory and imagination. Second, the question is asked, "How would we ever know that we were remembering (e.g.) the '4b' *correctly*, unless we did have a trace or a copy of the '4b' with which to compare the present memory?". For both of these reasons, our critic was unwilling to give up the belief that memory traces must exist. At the time, we did not examine these objections, claiming that such an examination would have been premature. Now we have completed our examination of trace theory. Having seen that the theory contains fundamental errors, and that the idea of a memory trace is essentially insubstantial, and even incoherent, our critic would, perhaps, be less inclined to maintain that memory traces must exist. But if we do reject the idea of the memory trace, what are we to make of these two objections? Our critic may now be imagined as saying, "Granted that your criticisms have destroyed the plausibility of memory traces, how in the world can we get along without them?".

Let us examine the first objection. Our critic asks, "If there is no trace, what would be the difference between someone remembering the '4b', and someone imagining it, inventing it anew, so

* See p. 38.

to speak? Without a trace, how could we distinguish between these two very different sorts of event? Take the case of someone humming the '4b'. It would not be a case of *remembering* (even if it looked like a case of remembering) unless we could follow a series of causal links back to the hearing of a tune. So the existence of memory traces is necessary in order to preserve (or make sense of) the distinction between memory and imagination."

In reply, it can be said that the distinction between memory and imagination does not rest on the supposed existence of some brain state or brain process. Let us consider an example. Suppose Mr. A is whistling something. Suppose that I don't know the tune, but I ask Mr. A what it is he's whistling. He answers, "It's Bach's lovely 'Fourth Brandenburg Concerto'." Let us suppose that Mr. A's answer was correct. Now, are we supposed to take seriously the suggestion that perhaps Mr. A wasn't *remembering* the '4b', but was 'inventing it anew'? It is obvious that our ability to distinguish between memory and invention does not depend on the existence or non-existence of some neurophysiological phenomenon. We wouldn't know how to begin to look for Mr. A's memory trace of '4b', even if there were such a thing. That surely doesn't prevent us from knowing, in the case envisioned, that Mr. A was whistling the tune from memory. (In contrast, imagine a case where he is whistling haltingly as he reads from a musical score. In this case we wouldn't say that he was whistling the tune from memory.) We have never before made use of traces in order to distinguish cases of remembering music from cases of inventing or creating new music; it would be odd if we suddenly developed this need.

But our critic may now say, "I admit that we don't actually rely on neurophysiological data in order to distinguish cases of remembering from cases of imagining. Nevertheless we must suppose that we *could*, in a bona fide case of remembering, trace a series of causal links back to the hearing of a tune. If there were no such causal links (i.e., no traces), then it just would not be a case of remembering." Clearly, someone who insists on this point feels that a case of (e.g.) humming the '4b' *could not possibly* be a case of remembering, unless there were a trace present. In other words, the existence of the 'step-by-step causal chain' is a *requirement* for memory. For such a person, a trace is one of the *defining features* of memory. And such an arbitrary requirement is no longer reasonable, in light of our examination of the idea of the trace. Of course, defining memory in this way would settle the issue of whether or not memory requires a trace. But then, if our attack on trace theory has been effective, the critic will be left with the consequence that there never have been any bona fide cases of remembering. Hopefully, this consequence would dissuade our critic from insisting that memory requires a trace by definition.

Let us now turn to the critic's second objection. "How would we ever know that we were remembering the '4b' *correctly*, unless we *did* have a copy, a trace, of the '4b' with which to compare the present memory?" I believe that this question, or one like it, will have occurred to many readers of this work. The question points to a problem of deep importance in philosophy. There is a very strong and common urge to say that, in order to recognize something '*x*', one must possess a copy or image of '*x*'. It cannot be that (for example) I just recognize the '4b' straight off, without anything to guide me. In order to recognize this complex piece of music (or complex pattern of sounds) I must somehow have within me a copy, a representation, a trace of the '4b'. When I recognize the '4b', this feat is made possible by a process of comparing the actual music with my internal copy. Without such an internal copy, I would have nothing to guide me. Consider the following: "In order to learn to discriminate squares from triangles, a child must somehow acquire an internal representation of a square, and of a triangle." Here the same problem makes its appearance as an unverified, a priori description of the process of *learning*. When I recite the alphabet, how do I know that the 11th letter is 'k'? Surely I must somehow have within me a copy of the alphabet, which is in some sense 'read off' when I recite. Otherwise, how could we explain the fact that I make the right sounds in the right order? People may differ as to whether or not such an internal copy is an image, a neurophysiological structure (i.e., a trace) or something else. But surely it is unthinkable that I might be able to simply recognize something (or know how to recite something) without consulting some sort of copy. How could I *just know* that New York is north of New Orleans, without some form of map, or representation of the United States?

In the *Blue Book*, Wittgenstein makes an observation which should, I think, utterly destroy the temptation to insist that recognition requires a copy of the thing recognized:

> If I give someone the order, "Fetch me a red flower from that meadow", how is he to know what sort of flower to bring, as I have only given him a *word*? Now the answer that one might suggest first is that he went to look for a red flower carrying a red image in his mind, and comparing it with the flowers to see which of them had the color of the image. . . . But this is not the only way of searching, and it isn't the usual way. We go, look about us, walk up to a flower and pick it, without comparing it to anything. To see that the process of obeying the order can be of this kind, consider the order, "*imagine* a red patch". You are not tempted in this case to think that *before* obeying you must have imagined a red patch to serve you as a pattern for the red patch which you were ordered to imagine. (his italics)[19]

These remarks take place in a discussion on the topic of meaning, but their application to our problem is clear. If you cannot recognize the '4b' straight off, without comparing it to some copy or image, then how can you recognize your copy or image of the '4b'? Do you recognize *that* straight off? If you don't recognize the image straight off, then do you need a second image which you compare to the first image, in order to make sure that the first image is correct? (That, of course, leads to an infinite regress of images or copies.) On the other hand, if you *can* recognize your image of the '4b' straight off, then why can't you recognize the '4b' straight off?

Wittgenstein is right; we must reject the idea that, in order to recognize something, one must possess an image or copy of the thing recognized. Along with this, we should reject the idea that memory requires a trace. To our critic's puzzled question, "How, without a trace to compare it with, can I know that I am whistling the '4b' correctly?", we may reply as follows. However strongly you may be tempted to call upon the idea of a trace or copy, it cannot help you. If we cannot recognize the '4b' straight off, then we cannot recognize the copy straight off, either. If we cannot know that we are whistling the '4b' correctly, in the absence of a trace to compare it with, then we also cannot know that the trace (or image) is the correct one – without a second image to compare it with (and so on, infinitely). Or is it that mental images have something about them which makes it possible for us to recognize them straight off? (The mental picture can do what no ordinary picture can do.) Wittgenstein says, "It was in fact just the occult character of the mental process which you needed for your purposes".[20] No, despite our critic's objections, we must abandon the idea of the memory trace. Despite first appearances, trace theory does not offer a scientific account of human memory.

This conclusion really does seem to leave us in a bind. If trace theory is the only possible mechanistic explanation of memory, and if trace theory commits us to the nonsensical idea of a magical mechanism, then we seem to have reached an impasse. It looks like any attempt to provide a scientific account of memory must fail. And that conclusion must strike terror, or at least confusion into the hearts of many people who have been brought up in our culture. After all, if the only possible scientific attempt to explain memory leaves us commited to the existence of a magical mechanism, then it seems as if science must go out the window. How many scientists believe in magic? If there is no scientific explanation of the phenomena of memory, then it seems as if memory is something beyond science – something supernatural, something magical. But for many people, the existence of the supernatural cannot be seriously contemplated. If there is even one supernatural or magical phenomenon, then the universe is a supernatural, magical place. We might as well start believing in Santa Claus, in fairies and angels, in miracles. So the question must be asked, "Is that what this book comes to?". Am I trying to convince people that memory is a supernatural phenomenon? After all, I *have* tried to show that there cannot be a scientific, mechanistic, causal explanation for memory.

The answer is, No, I do not believe that memory is a supernatural phenomenon. I do not believe in the supernatural, in ghosts, in magic, in fairies, in Santa Claus. But how can this be? If I am saying that memory is not susceptible of scientific explanation, then mustn't I be saying that memory is supernatural? I do not believe so. I believe that such a conclusion is based on a confusion. Specifically, it seems to be the product of a specious dichotomy. The assumption is this: Either memory can be explained scientifically, or else it is a supernatural phenomenon. This dichotomy is a particular example of a more general dichotomy: All phenomena are either natural phenomena, subject to scientific explanation, or else they are not natural, but supernatural phenomena. I can think of two ways of pinpointing the mistake in this reasoning. The first way is to say that it contains a mistaken assumption. The assumption is that all natural phenomena must have a scientific explanation. The second way of pinpointing the mistake is to say that the general dichotomy mentioned above is supposed to be exhaustive, but isn't. The dichotomy, once again, is this: All phenomena are either susceptible of scientific explanation, or else they are supernatural. The dichotomy is not really exhaustive, because it doesn't include the following possibility: A phenomenon might not be susceptible of scientific explanation, yet still be perfectly ordinary. Who says that the only alternative to a scientific explanation is a magical or supernatural explanation? Or, alternatively, who says that there is a scientific explanation for every natural phenomenon? For these are the assumptions behind the idea that my reasoning forced us to a belief in the supernatural. And these assumptions are mistaken.

Think of the extremely complex phenomenon which we call 'marriage'. It is a social phenomenon. It involves social institutions, customs, personal relationships, biology, and all sorts of things. Yet it seems to be a real, almost tangible phenomenon. I mean, for example, that we can talk about a particular marriage, and can even count marriages. My own belief is that there is no such thing as a scientific explanation or theory of marriage. In fact, I believe that the notion that there could be a scientific explanation of marriage is ludicrous. Now, do not mistake my intent. I am not about to offer a proof that there is no scientific explanation of marriage. I am aware that there are social scientists who believe that eventually all aspects of human behavior will be explained scientifically. But suppose that there is no scientific theory which explains marriage. Would this lead us to conclude that marriage is a supernatural phenomenon? Could it not just be that marriage is an enormously complex human institution? There is a scientific (biological) explanation of sex in the biological sense. But does this explanation constitute an explanation of love, for example? Insofar as biological sex has a scientific explanation, then marriage can be partially explained (or understood) in terms of the biological theory which describes sexual reproduction. But there is more to love than sexual reproduction, and more to marriage than just love. There is money, for example. Economic conditions can enter into an understanding of marriage. (Marrying someone for their money.) Is economics a science? And, more to the point, if economics is not a science, then is it therefore a study of the supernatural? I think there are all sorts of perfectly ordinary phenomena which are not susceptible of scientific explanation. This does not mean that they are supernatural or magical phenomena. Aren't there all sorts of explanations which are not scientific explanations? For example, in our culture, the bride (usually) wears white; mourners (usually) wear black. An explanation, for us, might be as follows: White, in our culture, is the symbol of untouched purity; black is regarded as the somber color of grief (or perhaps it symbolizes the dark never-ending night of death). Are these explanations scientific? Perhaps there is a scientific explanation underlying our use of symbols, and perhaps not. (I think not.) My point is, that if there is no scientific explanation of why white symbolizes purity to us, this would not show that our use of symbols is something supernatural. All it would show is that there are perfectly ordinary phenomena which do not have a scientific explanation.

But what is memory? If it doesn't have a scientific explanation, then what sort of phenomenon is it? I do not pretend to offer an explanation of memory. For one thing, I do not think that there is any one, unified phenomenon corresponding to the noun 'memory'. I believe that the phenomena of memory are extremely various. They form a *family*, in the sense referred to a few pages back. Of course there are aspects of memory which do have scientific explanations. For example, if my optic nerves had been severed at birth, I would never have seen anything. This would of course make it impossible for me to remember how anything looks, since I would never have *seen* how anything looks. And there is a science which explains the workings of the optic nerve. And there are probably areas of the brain which, if destroyed, would make it impossible for me to visualize anything. And such areas could be scientifically described (neuroanatomy). But it doesn't follow from this that there is a scientific explanation for memory. For example, there might be no scientific explanation of the fact that certain events are remembered while others are not. Here is an example. Suppose that Mr. A remembers that his wife wore a blue dress on the day they first met. (thirty years earlier). There may be an explanation for the fact that he remembers this detail over such a long time span. For example, he might tell us that he remembers that detail because he fell in love with his wife the moment he saw her. "How could I ever forget what she looked like on that day?" And if Mr. A's romantic explanation turns out to be the final word (i.e., if there is no *scientific* explanation of why that particular memory is so strong) does that mean that memory must be supernatural? That would only be so if we made the mistaken assumption that memory is either scientifically explicable, or else is supernatural, and magical.

All this is intended to convince the reader that there need not be a scientific, mechanistic, causal theory of memory. Magic is not the only alternative to science. The fact is that it was the attempt to provide a scientific explanation for memory which committed *the trace theorist* to the existence of a magical mechanism. But we, on the other hand, have no such commitment. In fact, the shoe is on the other foot. We said that any *machine* which had to do the things which a mechanistic memory theory requires would have to be a magical machine. That is, any mechanism which is supposed to do what people do would be a magical machine. But who says that people are machines? It is the trace theorist who makes this assumption, not we. Paradoxically, it is the trace theorist's attempt to extend scientific explanation beyond its proper bounds which commits him to magic. We, on the other hand, do not assume that there must be a scientific explanation for *everything*. We do not assume that people are machines.

NOTES

[1] Cornford translation in *Plato: Collected Dialogues*, ed. by Edith Hamilton and Huntington Cairns, 1961, Pantheon Books, 1743 pp. p. 897.

[2] Aquinas, *Summa Theologica*, Paris Prima, QLXXVIII, art. IV.

[3] From *Models of Human Memory, op. cit.*, article by D. A. Norman and D. Rumelhart.

[4] *Models of Human Memory, op. cit.*, article cit., 'A system for perception and memory', pp. 19–20.

[5] Hobbes, *Leviathan*, E. P. Dutton & Co., New York, 1950, 630 pp., part I, Chapter 2, p. 8.

[6] *Ibid.*, p. 10.

[7] *Ibid.*, p. 11, his italics.

[8] See, for example, the issue of *Time* magazine for February 20, 1978.

[9] *The Neuropsychology of Lashley, op. cit.*, p. 501.

[10] See Wittgenstein's *Philosophical Investigations, op. cit.*, par. 47–8.
[11] W. Köhler, *The Task of Gestalt Psychology, op. cit.*, p. 122.
[12] *Ibid.*, p. 123.
[13] *The Neuropsychology of Lashley, op. cit.*, p. 535.
[14] *Models of Human Memory, op. cit.*, article cit., 'A system for perception and memory', p. 22.
[15] By Jerry A. Fodor, 165 pp., Random House, New York, 1968.
[16] *Psychological Explanation, op. cit.*, pp. 25–6.
[17] The reader should not gather from this the mistaken notion that there is a set, 'X', which is the set of ashtrays. It is not fixed, immutably and for all time, what things are, and what things are not ashtrays.
[18] For a discussion of the idea of family resemblance, see Wittgenstein's *Philosophical Investigations* (*op. cit.*), see especially, Part I, paragraphs 65–71.
[19] Ludwig Wittgenstein, *The Blue and Brown Books*, Harper Torchbooks edition, p. 3.
[20] *Blue Book*, p. 5.

BIBLIOGRAPHY

Abbott, W. and Fields, W. S. (eds.), *Information Storage and Neural Control.* Springfield: Charles C. Thomas, 1963.

Ashby, W. R., *Design For A Brain.* 2nd ed., rev. John Wiley and Sons, Inc., 1960.

Broad, C. D., *Mind and Its Place in Nature.* New York: Harcourt, Brace & Comp. Inc., 1925, 674 pp.

Broadbent, D. E. and Pribram, K. H. (eds.), *Biology of Memory.* New York and London: Academic Press, 1970, 323 pp.

Burns, B. D., *The Mammalian Cerebral Cortex*, 1958.

Culbertson, J. T., *Consciousness and Behavior.* Dubuque: W. C. Brown, 1950, 210 pp.

Eccles, Sir John, *Neurophysiological Basis of Mind.* Oxford, 1952.

Elliott, H. Chandler, *The Shape of Intelligence.* New York: Scribner's Sons, 1969.

Feigl, H., Scriven, M., and Maxwell, G., 'The 'Mental' and the 'Physical',' *Minnesota Studies in the Philosophy of Science*, Vol. II (1958).

Fodor, J. A., *Psychological Explanation.* New York: Random House, 1968, 165 pp.

Harris, T. A., *I'm O.K. – You're O.K.* New York: Harper and Row, 1967, 280 pp.

Hobbes, Thomas, *Leviathan.* New York: E. P. Dutton & Co., 1950, 630 pp.

Köhler, Wolfgang, *Gestalt Psychology.* New York: H. Liveright, 1929, 403 pp.

Köhler, Wolfgang, *The Task of Gestalt Psychology.* Princeton: Princeton University Press, 1969, 164 pp.

Lashley, K. S., *The Neuropsychology of Lashley.* Ed. by F. A. Beach and D. O. Hebb.

Malcolm, Norman, 'Conceivability of Mechanism,' *The Philosophical Review* (1968).

Malcolm, Norman, *Problems of Mind: Descartes to Wittgenstein.* New York: Harper and Row, 1971, 103 pp.

Munsat, S., *The Concept of Memory.* New York: Random House, 1967, 130 pp.

Norman, D. A. (ed.), *Models of Human Memory.* Academic Press, 1970, 537 pp.

Plato, *Collected Dialogues.* Ed. by Edith Hamilton and Huntington Cairns. Cornford's translation of *Theaetetus.* Pantheon Books, 1961, 1743 pp.

Pribram, K. H. (ed.), *Brain and Behavior 3: Memory Mechanisms*. Penguin Books, 1969, 524 pp.

Russell, B., *Analysis of Mind*. New York: Macmillan Co., 1921, 310 pp.

Russell, B., *The Problems of Philosophy*. New York: H. Holt & Co., 1921.

Russell, W. R., *Brain Memory Learning*. Oxford Press, 1959.

Ryle, G., *The Concept of Mind*. Hutchinson, 1949.

Skinner, B. F., *The Behavior of Organisms*. New York: Appleton-Century-Crofts, Inc., 1936, 457 pp.

Wilkie, J. S., *The Science of Mind and Brain*. London: Hutchinson House, 1953, 160 pp.

Wittgenstein, Ludwig, *The Blue and Brown Books*. Preliminary Studies for the *Philosophical Investigations*. New York: Harper and Row, 1965, 192 pp.

Wittgenstein, Ludwig, *On Certainty*. Ed. by G. E. M. Anscombe and G. H. von Wright. Trans. by Denis Paul and G. E. M. Anscombe. New York: J. & J. Harper Editions, 1969, 90 pp.

Wittgenstein, Ludwig, *Philosophical Investigations*. Trans. by G. E. M. Anscombe. 2nd ed. New York: Macmillan Co., 1958, 232 pp.

Wittgenstein, Ludwig, *Zettel*. Ed. by G. E. M. Anscombe and G. H. von Wright. Trans. by G. E. M. Anscombe. Berkeley: University of California Press, 1967, 124 pp.

INDEX OF NAMES

INDEX OF SUBJECTS

SYNTHESE LIBRARY

Studies in Epistemology, Logic, Methodology,
and Philosophy of Science

1. J. M. Bocheński, *A Precis of Mathematical Logic.* 1959, X + 100 pp.
2. P. L. Guiraud, *Problèmes et méthodes de la statistique linguistique.* 1960, VI + 146 pp.
3. Hans Freudenthal (ed.), *The Concept and the Role of the Model in Mathematics and Natural and Social Sciences. Proceedings of a Colloquium held at Utrecht, The Netherlands, January 1960.* 1961, VI + 194 pp.
4. Evert W. Beth, *Formal Methods. An Introduction to Symbolic Logic and the Study of Effective Operations in Arithmetic and Logic.* 1962, XIV + 170 pp.
5. B. H. Kazemier and D. Vuysje (eds.), *Logic and Language. Studies Dedicated to Professor Rudolf Carnap on the Occasion of His Seventieth Birthday.* 1962, VI + 256 pp.
6. Marx W. Wartofsky (ed.), *Proceedings of the Boston Colloquium for the Philosophy of Science 1961-1962,* Boston Studies in the Philosophy of Science (ed. by Robert S. Cohen and Marx W. Wartofsky), Volume I. 1963, VIII + 212 pp.
7. A. A. Zinov'ev, *Philosophical Problems of Many-Valued Logic.* 1963, XIV + 155 pp.
8. Georges Gurvitch, *The Spectrum of Social Time.* 1964, XXVI + 152 pp.
9. Paul Lorenzen, *Formal Logic.* 1965, VIII + 123 pp.
10. Robert S. Cohen and Marx W. Wartofsky (eds.), *In Honor of Philipp Frank,* Boston Studies in the Philosophy of Science (ed. by Robert S. Cohen and Marx W. Wartofsky), Volume II. 1965, XXXIV + 475 pp.
11. Evert W. Beth, *Mathematical Thought. An Introduction to the Philosophy of Mathematics.* 1965, XII + 208 pp.
12. Evert W. Beth and Jean Piaget, *Mathematical Epistemology and Psychology.* 1966, XII + 326 pp.
13. Guido Küng, *Ontology and the Logistic Analysis of Language. An Enquiry into the Contemporary Views on Universals.* 1967, XI + 210 pp.
14. Robert S. Cohen and Marx W. Wartofsky (eds.), *Proceedings of the Boston Colloquium for the Philosophy of Science 1964-1966, in Memory of Norwood Russell Hanson,* Boston Studies in the Philosophy of Science (ed. by Robert S. Cohen and Marx W. Wartofsky), Volume III. 1967, XLIX + 489 pp.

15. C. D. Broad, *Induction, Probability, and Causation. Selected Papers.* 1968, XI + 296 pp.
16. Günther Patzig, *Aristotle's Theory of the Syllogism. A Logical-Philosophical Study of Book A of the Prior Analytics.* 1968, XVII + 215 pp.
17. Nicholas Rescher, *Topics in Philosophical Logic.* 1968, XIV + 347 pp.
18. Robert S. Cohen and Marx W. Wartofsky (eds.), *Proceedings of the Boston Colloquium for the Philosophy of Science 1966-1968,* Boston Studies in the Philosophy of Science (ed. by Robert S. Cohen and Marx W. Wartofsky), Volume IV. 1969, VIII + 537 pp.
19. Robert S. Cohen and Marx W. Wartofsky (eds.), *Proceedings of the Boston Colloquium for the Philosophy of Science 1966-1968,* Boston Studies in the Philosophy of Science (ed. by Robert S. Cohen and Marx W. Wartofsky), Volume V. 1969, VIII + 482 pp.
20. J.W. Davis, D. J. Hockney, and W. K. Wilson (eds.), *Philosophical Logic.* 1969, VIII + 277 pp.
21. D. Davidson and J. Hintikka (eds.), *Words and Objections: Essays on the Work of W.V. Quine.* 1969, VIII + 366 pp.
22. Patrick Suppes, *Studies in the Methodology and Foundations of Science. Selected Papers from 1911 to 1969.* 1969, XII + 473 pp.
23. Jaakko Hintikka, *Models for Modalities. Selected Essays.* 1969, IX + 220 pp.
24. Nicholas Rescher *et al.* (eds.), *Essays in Honor of Carl G. Hempel. A Tribute on the Occasion of His Sixty-Fifth Birthday.* 1969, VII + 272 pp.
25. P. V. Tavanec (ed.), *Problems of the Logic of Scientific Knowledge.* 1969, XII + 429 pp.
26. Marshall Swain (ed.), *Induction, Acceptance, and Rational Belief.* 1970, VII + 232 pp.
27. Robert S. Cohen and Raymond J. Seeger (eds.), *Ernst Mach: Physicist and Philosopher,* Boston Studies in the Philosophy of Science (ed. by Robert S. Cohen and Marx W. Wartofsky), Volume VI. 1970, VIII + 295 pp.
28. Jaakko Hintikka and Patrick Suppes, *Information and Inference.* 1970, X + 336 pp.
29. Karel Lambert, *Philosophical Problems in Logic. Some Recent Developments.* 1970, VII + 176 pp.
30. Rolf A. Eberle, *Nominalistic Systems.* 1970, IX + 217 pp.
31. Paul Weingartner and Gerhard Zecha (eds.), *Induction, Physics, and Ethics: Proceedings and Discussions of the 1968 Salzburg Colloquium in the Philosophy of Science.* 1970, X + 382 pp.
32. Evert W. Beth, *Aspects of Modern Logic.* 1970, XI + 176 pp.
33. Risto Hilpinen (ed.), *Deontic Logic: Introductory and Systematic Readings.* 1971, VII + 182 pp.
34. Jean-Louis Krivine, *Introduction to Axiomatic Set Theory.* 1971, VII + 98 pp.
35. Joseph D. Sneed, *The Logical Structure of Mathematical Physics.* 1971, XV + 311 pp.
36. Carl R. Kordig, *The Justification of Scientific Change.* 1971, XIV + 119 pp.
37. Milič Čapek, *Bergson and Modern Physics,* Boston Studies in the Philosophy of Science (ed. by Robert S. Cohen and Marx W. Wartofsky), Volume VII. 1971, XV + 414 pp.

38. Norwood Russell Hanson, *What I Do Not Believe, and Other Essays* (ed. by Stephen Toulmin and Harry Woolf), 1971, XII + 390 pp.
39. Roger C. Buck and Robert S. Cohen (eds.), *PSA 1970. In Memory of Rudolf Carnap*, Boston Studies in the Philosophy of Science (ed. by Robert S. Cohen and Marx W. Wartofsky), Volume VIII. 1971, LXVI + 615 pp. Also available as paperback.
40. Donald Davidson and Gilbert Harman (eds.), *Semantics of Natural Language.* 1972, X + 769 pp. Also available as paperback.
41. Yehoshua Bar-Hillel (ed.), *Pragmatics of Natural Languages.* 1971, VII + 231 pp.
42. Sören Stenlund, *Combinators, λ-Terms and Proof Theory.* 1972, 184 pp.
43. Martin Strauss, *Modern Physics and Its Philosophy. Selected Papers in the Logic, History, and Philosophy of Science.* 1972, X + 297 pp.
44. Mario Bunge, *Method, Model and Matter.* 1973, VII + 196 pp.
45. Mario Bunge, *Philosophy of Physics.* 1973, IX + 248 pp.
46. A. A. Zinov'ev, *Foundations of the Logical Theory of Scientific Knowledge (Complex Logic)*, Boston Studies in the Philosophy of Science (ed. by Robert S. Cohen and Marx W. Wartofsky), Volume IX. Revised and enlarged English edition with an appendix, by G. A. Smirnov, E. A. Sidorenka, A. M. Fedina, and L. A. Bobrova. 1973, XXII + 301 pp. Also available as paperback.
47. Ladislav Tondl, *Scientific Procedures*, Boston Studies in the Philosophy of Science (ed. by Robert S. Cohen and Marx W. Wartofsky), Volume X. 1973, XII + 268 pp. Also available as paperback.
48. Norwood Russell Hanson, *Constellations and Conjectures* (ed. by Willard C. Humphreys, Jr.). 1973, X + 282 pp.
49. K. J. J. Hintikka, J. M. E. Moravcsik, and P. Suppes (eds.), *Approaches to Natural Language. Proceedings of the 1970 Stanford Workshop on Grammar and Semantics.* 1973, VIII + 526 pp. Also available as paperback.
50. Mario Bunge (ed.), *Exact Philosophy – Problems, Tools, and Goals.* 1973, X + 214 pp.
51. Radu J. Bogdan and Ilkka Niiniluoto (eds.), *Logic, Language, and Probability. A Selection of Papers Contributed to Sections IV, VI, and XI of the Fourth International Congress for Logic, Methodology, and Philosophy of Science, Bucharest, September 1971.* 1973, X + 323 pp.
52. Glenn Pearce and Patrick Maynard (eds.), *Conceptual Change.* 1973, XII + 282 pp.
53. Ilkka Niiniluoto and Raimo Tuomela, *Theoretical Concepts and Hypothetico-Inductive Inference.* 1973, VII + 264 pp.
54. Roland Fraïssé, *Course of Mathematical Logic* – Volume 1: *Relation and Logical Formula.* 1973, XVI + 186 pp. Also available as paperback.
55. Adolf Grünbaum, *Philosophical Problems of Space and Time.* Second, enlarged edition, Boston Studies in the Philosophy of Science (ed. by Robert S. Cohen and Marx W. Wartofsky), Volume XII. 1973, XXIII + 884 pp. Also available as paperback.
56. Patrick Suppes (ed.), *Space, Time, and Geometry.* 1973, XI + 424 pp.
57. Hans Kelsen, *Essays in Legal and Moral Philosophy*, selected and introduced by Ota Weinberger. 1973, XXVIII + 300 pp.
58. R. J. Seeger and Robert S. Cohen (eds.), *Philosophical Foundations of Science. Proceedings of an AAAS Program, 1969*, Boston Studies in the Philosophy of

Science (ed. by Robert S. Cohen and Marx W. Wartofsky), Volume XI. 1974, X + 545 pp. Also available as paperback.

59. Robert S. Cohen and Marx W. Wartofsky (eds.), *Logical and Epistemological Studies in Contemporary Physics,* Boston Studies in the Philosophy of Science (ed. by Robert S. Cohen and Marx W. Wartofsky), Volume XIII. 1973, VIII + 462 pp. Also available as paperback.
60. Robert S. Cohen and Marx W. Wartofsky (eds.), *Methodological and Historical Essays in the Natural and Social Sciences. Proceedings of the Boston Colloquium for the Philosophy of Science 1969-1972,* Boston Studies in the Philosophy of Science (ed. by Robert S. Cohen and Marx W. Wartofsky), Volume XIV. 1974, VIII + 405 pp. Also available as paperback.
61. Robert S. Cohen, J. J. Stachel and Marx W. Wartofsky (eds.), *For Dirk Struik. Scientific, Historical and Political Essays in Honor of Dirk J. Struik,* Boston Studies in the Philosophy of Science (ed. by Robert S. Cohen and Marx W. Wartofsky), Volume XV. 1974, XXVII + 652 pp. Also available as paperback.
62. Kazimierz Ajdukiewicz, *Pragmatic Logic,* transl. from the Polish by Olgierd Wojtasiewicz. 1974, XV + 460 pp.
63. Sören Stenlund (ed.), *Logical Theory and Semantic Analysis. Essays Dedicated to Stig Kanger on His Fiftieth Birthday.* 1974, V + 217 pp.
64. Kenneth F. Schaffner and Robert S. Cohen (eds.), *Proceedings of the 1972 Biennial Meeting, Philosophy of Science Association,* Boston Studies in the Philosophy of Science (ed. by Robert S. Cohen and Marx W. Wartofsky), Volume XX. 1974, IX + 444 pp. Also available as paperback.
65. Henry E. Kyburg, Jr., *The Logical Foundations of Statistical Inference.* 1974, IX + 421 pp.
66. Marjorie Grene, *The Understanding of Nature: Essays in the Philosophy of Biology,* Boston Studies in the Philosophy of Science (ed. by Robert S. Cohen and Marx W. Wartofsky), Volume XXIII. 1974, XII + 360 pp. Also available as paperback.
67. Jan M. Broekman, *Structuralism: Moscow, Prague, Paris.* 1974, IX + 117 pp.
68. Norman Geschwind, *Selected Papers on Language and the Brain,* Boston Studies in the Philosophy of Science (ed. by Robert S. Cohen and Marx W. Wartofsky), Volume XVI. 1974, XII + 549 pp. Also available as paperback.
69. Roland Fraïssé, *Course of Mathematical Logic* – Volume 2: *Model Theory.* 1974, XIX + 192 pp.
70. Andrzej Grzegorczyk, *An Outline of Mathematical Logic. Fundamental Results and Notions Explained with All Details.* 1974, X + 596 pp.
71. Franz von Kutschera, *Philosophy of Language.* 1975, VII + 305 pp.
72. Juha Manninen and Raimo Tuomela (eds.), *Essays on Explanation and Understanding. Studies in the Foundations of Humanities and Social Sciences.* 1976, VII + 440 pp.
73. Jaakko Hintikka (ed.), *Rudolf Carnap, Logical Empiricist. Materials and Perspectives.* 1975, LXVIII + 400 pp.
74. Milič Čapek (ed.), *The Concepts of Space and Time. Their Structure and Their Development,* Boston Studies in the Philosophy of Science (ed. by Robert S. Cohen and Marx W. Wartofsky), Volume XXII. 1976, LVI + 570 pp. Also available as paperback.

75. Jaakko Hintikka and Unto Remes, *The Method of Analysis. Its Geometrical Origin and Its General Significance,* Boston Studies in the Philosophy of Science (ed. by Robert S. Cohen and Marx W. Wartofsky), Volume XXV. 1974, XVIII + 144 pp. Also available as paperback.
76. John Emery Murdoch and Edith Dudley Sylla, *The Cultural Context of Medieval Learning. Proceedings of the First International Colloquium on Philosophy, Science, and Theology in the Middle Ages – September 1973,* Boston Studies in the Philosophy of Science (ed. by Robert S. Cohen and Marx W. Wartofsky), Volume XXVI. 1975, X + 566 pp. Also available as paperback.
77. Stefan Amsterdamski, *Between Experience and Metaphysics. Philosophical Problems of the Evolution of Science,* Boston Studies in the Philosophy of Science (ed. by Robert S. Cohen and Marx W. Wartofsky), Volume XXXV. 1975, XVIII + 193 pp. Also available as paperback.
78. Patrick Suppes (ed.), *Logic and Probability in Quantum Mechanics.* 1976, XV + 541 pp.
79. Hermann von Helmholtz: *Epistemological Writings. The Paul Hertz/Moritz Schlick Centenary Edition of 1921 with Notes and Commentary by the Editors.* (Newly translated by Malcolm F. Lowe. Edited with an Introduction and Bibliography, by Robert S. Cohen and Yehuda Elkana), Boston Studies in the Philosophy of Science (ed. by Robert S. Cohen and Marx W. Wartofsky), Volume XXXVII. 1977, XXXVIII+204 pp. Also available as paperback.
80. Joseph Agassi, *Science in Flux,* Boston Studies in the Philosophy of Science (ed. by Robert S. Cohen and Marx W. Wartofsky), Volume XXVIII. 1975, XXVI + 553 pp. Also available as paperback.
81. Sandra G. Harding (ed.), *Can Theories Be Refuted? Essays on the Duhem-Quine Thesis.* 1976, XXI + 318 pp. Also available as paperback.
82. Stefan Nowak, *Methodology of Sociological Research: General Problems.* 1977, XVIII + 504 pp.
83. Jean Piaget, Jean-Blaise Grize, Alina Szeminska, and Vinh Bang, *Epistemology and Psychology of Functions,* Studies in Genetic Epistemology, Volume XXIII. 1977, XIV+205 pp.
84. Marjorie Grene and Everett Mendelsohn (eds.), *Topics in the Philosophy of Biology,* Boston Studies in the Philosophy of Science (ed. by Robert S. Cohen and Marx W. Wartofsky), Volume XXVII. 1976, XIII + 454 pp. Also available as paperback.
85. E. Fischbein, *The Intuitive Sources of Probabilistic Thinking in Children.* 1975, XIII + 204 pp.
86. Ernest W. Adams, *The Logic of Conditionals. An Application of Probability to Deductive Logic.* 1975, XIII + 156 pp.
87. Marian Przełęcki and Ryszard Wójcicki (eds.), *Twenty-Five Years of Logical Methodology in Poland.* 1977, VIII + 803 pp.
88. J. Topolski, *The Methodology of History.* 1976, X + 673 pp.
89. A. Kasher (ed.), *Language in Focus: Foundations, Methods and Systems. Essays Dedicated to Yehoshua Bar-Hillel,* Boston Studies in the Philosophy of Science (ed. by Robert S. Cohen and Marx W. Wartofsky), Volume XLIII. 1976, XXVIII + 679 pp. Also available as paperback.
90. Jaakko Hintikka, *The Intentions of Intentionality and Other New Models for Modalities.* 1975, XVIII + 262 pp. Also available as paperback.

91. Wolfgang Stegmüller, *Collected Papers on Epistemology, Philosophy of Science and History of Philosophy*, 2 Volumes, 1977, XXVII + 525 pp.
92. Dov M. Gabbay, *Investigations in Modal and Tense Logics with Applications to Problems in Philosophy and Linguistics.* 1976, XI + 306 pp.
93. Radu J. Bogdan, *Local Induction.* 1976, XIV + 340 pp.
94. Stefan Nowak, *Understanding and Prediction: Essays in the Methodology of Social and Behavioral Theories.* 1976, XIX + 482 pp.
95. Peter Mittelstaedt, *Philosophical Problems of Modern Physics*, Boston Studies in the Philosophy of Science (ed. by Robert S. Cohen and Marx W. Wartofsky), Volume XVIII. 1976, X + 211 pp. Also available as paperback.
96. Gerald Holton and William Blanpied (eds.), *Science and Its Public: The Changing Relationship*, Boston Studies in the Philosophy of Science (ed. by Robert S. Cohen and Marx W. Wartofsky), Volume XXXIII. 1976, XXV + 289 pp. Also available as paperback.
97. Myles Brand and Douglas Walton (eds.), *Action Theory. Proceedings of the Winnipeg Conference on Human Action, Held at Winnipeg, Manitoba, Canada, 9-11 May 1975.* 1976, VI + 345 pp.
98. Risto Hilpinen, *Knowledge and Rational Belief.* 1979 (forthcoming).
99. R. S. Cohen, P. K. Feyerabend, and M. W. Wartofsky (eds.), *Essays in Memory of Imre Lakatos*, Boston Studies in the Philosophy of Science (ed. by Robert S. Cohen and Marx W. Wartofsky), Volume XXXIX. 1976, XI + 762 pp. Also available as paperback.
100. R. S. Cohen and J. J. Stachel (eds.), *Selected Papers of Léon Rosenfeld*, Boston Studies in the Philosophy of Science (ed. by Robert S. Cohen and Marx W. Wartofsky), Volume XXI. 1978, XXX + 927 pp.
101. R. S. Cohen, C. A. Hooker, A. C. Michalos, and J. W. van Evra (eds.), *PSA 1974: Proceedings of the 1974 Biennial Meeting of the Philosophy of Science Association*, Boston Studies in the Philosophy of Science (ed. by Robert S. Cohen and Marx W. Wartofsky), Volume XXXII. 1976, XIII + 734 pp. Also available as paperback.
102. Yehuda Fried and Joseph Agassi, *Paranoia: A Study in Diagnosis*, Boston Studies in the Philosophy of Science (ed. by Robert S. Cohen and Marx W. Wartofsky), Volume L. 1976, XV + 212 pp. Also available as paperback.
103. Marian Przełęcki, Klemens Szaniawski, and Ryszard Wójcicki (eds.), *Formal Methods in the Methodology of Empirical Sciences.* 1976, 455 pp.
104. John M. Vickers, *Belief and Probability.* 1976, VIII + 202 pp.
105. Kurt H. Wolff, *Surrender and Catch: Experience and Inquiry Today*, Boston Studies in the Philosophy of Science (ed. by Robert S. Cohen and Marx W. Wartofsky), Volume LI. 1976, XII + 410 pp. Also available as paperback.
106. Karel Kosík, *Dialectics of the Concrete*, Boston Studies in the Philosophy of Science (ed. by Robert S. Cohen and Marx W. Wartofsky), Volume LII. 1976, VIII + 158 pp. Also available as paperback.
107. Nelson Goodman, *The Structure of Appearance*, Boston Studies in the Philosophy of Science (ed. by Robert S. Cohen and Marx W. Wartofsky), Volume LIII. 1977, L + 285 pp.
108. Jerzy Giedymin (ed.), *Kazimierz Ajdukiewicz: The Scientific World-Perspective and Other Essays, 1931 - 1963.* 1978, LIII + 378 pp.

109. Robert L. Causey, *Unity of Science*. 1977, VIII+185 pp.
110. Richard E. Grandy, *Advanced Logic for Applications*. 1977, XIV + 168 pp.
111. Robert P. McArthur, *Tense Logic*. 1976, VII + 84 pp.
112. Lars Lindahl, *Position and Change: A Study in Law and Logic*. 1977, IX + 299 pp.
113. Raimo Tuomela, *Dispositions*. 1978, X + 450 pp.
114. Herbert A. Simon, *Models of Discovery and Other Topics in the Methods of Science*, Boston Studies in the Philosophy of Science (ed. by Robert S. Cohen and Marx W. Wartofsky), Volume LIV. 1977, XX + 456 pp. Also available as paperback.
115. Roger D. Rosenkrantz, *Inference, Method and Decision*. 1977, XVI + 262 pp. Also available as paperback.
116. Raimo Tuomela, *Human Action and Its Explanation. A Study on the Philosophical Foundations of Psychology*. 1977, XII + 426 pp.
117. Morris Lazerowitz, *The Language of Philosophy. Freud and Wittgenstein*, Boston Studies in the Philosophy of Science (ed. by Robert S. Cohen and Marx W. Wartofsky), Volume LV. 1977, XVI + 209 pp.
118. Tran Duc Thao, *Origins of Language and Consciousness*, Boston Studies in the Philosophy of Science (ed. by Robert S. Cohen and Marx. W. Wartofsky), Volume LVI. 1979 (forthcoming).
119. Jerzy Pelč, *Semiotics in Poland, 1894 – 1969*. 1977, XXVI + 504 pp.
120. Ingmar Pörn, *Action Theory and Social Science. Some Formal Models*. 1977, X + 129 pp.
121. Joseph Margolis, *Persons and Minds, The Prospects of Nonreductive Materialism*, Boston Studies in the Philosophy of Science (ed. by Robert S. Cohen and Marx W. Wartofsky), Volume LVII. 1977, XIV + 282 pp. Also available as paperback.
122. Jaakko Hintikka, Ilkka Niiniluoto, and Esa Saarinen (eds.), *Essays on Mathematical and Philosophical Logic. Proceedings of the Fourth Scandinavian Logic Symposium and of the First Soviet-Finnish Logic Conference, Jyväskylä, Finland, 1976*. 1978, VIII + 458 pp. + index.
123. Theo A. F. Kuipers, *Studies in Inductive Probability and Rational Expectation*. 1978, XII + 145 pp.
124. Esa Saarinen, Risto Hilpinen, Ilkka Niiniluoto, and Merrill Provence Hintikka (eds.), *Essays in Honour of Jaakko Hintikka on the Occasion of His Fiftieth Birthday*. 1978, IX + 378 pp. + index.
125. Gerard Radnitzky and Gunnar Andersson (eds.), *Progress and Rationality in Science*, Boston Studies in the Philosophy of Science (ed. by Robert S. Cohen and Marx W. Wartofsky), Volume LVIII. 1978, X + 400 pp. + index. Also available as paperback.
126. Peter Mittelstaedt, *Quantum Logic*. 1978, IX + 149 pp.
127. Kenneth A. Bowen, *Model Theory for Modal Logic. Kripke Models for Modal Predicate Calculi*. 1978, X + 128 pp.
128. Howard Alexander Bursen, *Dismantling the Memory Machine. A Philosophical Investigation of Machine Theories of Memory*. 1978, XIII + 157 pp.
129. Marx W. Wartofsky, *Models: Representation and Scientific Understanding*, Boston Studies in the Philosophy of Science (ed. by Robert S. Cohen and Marx W. Wartofsky), Volume XLVIII. 1979 (forthcoming). Also available as a paperback.
130. Don Ihde, *Technics and Praxis. A Philosophy of Technology*, Boston Studies in

the Philosophy of Science (ed. by Robert S. Cohen and Marx W. Wartofsky), Volume XXIV. 1979 (forthcoming). Also available as a paperback.

131. Jerzy J. Wiatr (ed.), *Polish Essays in the Methodology of the Social Sciences*, Boston Studies in the Philosophy of Science (ed. by Robert S. Cohen and Marx W. Wartofsky), Volume XXIX. 1979 (forthcoming). Also available as a paperback.
132. Wesley C. Salmon (ed.), *Hans Reichenbach: Logical Empiricist.* 1979 (forthcoming).
133. R.-P. Horstmann and L. Krüger (eds.), *Transcendental Arguments and Science.* 1979 (forthcoming). Also available as a paperback.

SYNTHESE HISTORICAL LIBRARY

Texts and Studies
in the History of Logic and Philosophy

Editors:

1. M. T. Beonio-Brocchieri Fumagalli, *The Logic of Abelard.* Translated from the Italian. 1969, IX + 101 pp.
2. Gottfried Wilhelm Leibniz, *Philosophical Papers and Letters.* A selection translated and edited, with an introduction, by Leroy E. Loemker. 1969, XII + 736 pp.
3. Ernst Mally, *Logische Schriften,* ed. by Karl Wolf and Paul Weingartner. 1971, X + 340 pp.
4. Lewis White Beck (ed.), *Proceedings of the Third International Kant Congress.* 1972, XI + 718 pp.
5. Bernard Bolzano, *Theory of Science,* ed. by Jan Berg. 1973, XV + 398 pp.
6. J. M. E. Moravcsik (ed.), *Patterns in Plato's Thought. Papers Arising Out of the 1971 West Coast Greek Philosophy Conference.* 1973, VIII + 212 pp.
7. Nabil Shehaby, *The Propositional Logic of Avicenna: A Translation from al-Shifā: al-Qiyās,* with Introduction, Commentary and Glossary. 1973, XIII + 296 pp.
8. Desmond Paul Henry, *Commentary on De Grammatico: The Historical-Logical Dimensions of a Dialogue of St. Anselm's.* 1974, IX + 345 pp.
9. John Corcoran, *Ancient Logic and Its Modern Interpretations.* 1974, X + 208 pp.
10. E. M. Barth, *The Logic of the Articles in Traditional Philosophy.* 1974, XXVII + 533 pp.
11. Jaakko Hintikka, *Knowledge and the Known. Historical Perspectives in Epistemology.* 1974, XII + 243 pp.
12. E. J. Ashworth, *Language and Logic in the Post-Medieval Period.* 1974, XIII + 304 pp.
13. Aristotle, *The Nicomachean Ethics.* Translated with Commentaries and Glossary by Hypocrates G. Apostle. 1975, XXI + 372 pp.
14. R. M. Dancy, *Sense and Contradiction: A Study in Aristotle.* 1975, XII + 184 pp.
15. Wilbur Richard Knorr, *The Evolution of the Euclidean Elements. A Study of the Theory of Incommensurable Magnitudes and Its Significance for Early Greek Geometry.* 1975, IX + 374 pp.
16. Augustine, *De Dialectica.* Translated with Introduction and Notes by B. Darrell Jackson. 1975, XI + 151 pp.

17. Arpád Szabó, *The Beginnings of Greek Mathematics.* 1979 (forthcoming).
18. Rita Guerlac, *Juan Luis Vives Against the Pseudodialecticians. A Humanist Attack on Medieval Logic.* Texts, with translation, introduction and notes. 1978, xiv + 227 pp. + index.

SYNTHESE LANGUAGE LIBRARY

Texts and Studies
in Linguistics and Philosophy

1. Henry Hiż (ed.), *Questions.* 1977, xvii + 366 pp.
2. William S. Cooper, *Foundations of Logico-Linguistics. A Unified Theory of Information, Language, and Logic.* 1978, xvi + 249 pp.
3. Avishai Margalit (ed.), *Meaning and Use. Papers Presented at the Second Jerusalem Philosophical Encounter, April 1976.* 1978 (forthcoming).
4. F. Guenthner and S. J. Schmidt (eds.), *Formal Semantics and Pragmatics for Natural Languages.* 1978, viii + 374 pp. + index.
5. Esa Saarinen (ed.), *Game-Theoretical Semantics.* 1978, xiv + 379 pp. + index.
6. F. J. Pelletier (ed.), *Mass Terms: Some Philosophical Problems.* 1978, xiv + 300 pp. + index.